The Nature Conservancy
Protecting nature. Preserving life.™

# COMMEMORATING
# 50 YEARS
## OF CONSERVATION HEROES
### IN WASHINGTON AND AROUND THE WORLD
## 2010

CONSERVING
50
YEARS
WASHINGTON

# JANE GOODALL
# 50 YEARS AT GOMBE

## A TRIBUTE TO FIVE DECADES OF WILDLIFE RESEARCH, EDUCATION, AND CONSERVATION

*Jane Goodall*

# JANE GOODALL
# 50 YEARS AT GOMBE

## A TRIBUTE TO FIVE DECADES OF WILDLIFE
## RESEARCH, EDUCATION, AND CONSERVATION

## Jane Goodall
with the JANE GOODALL INSTITUTE

Stewart, Tabori & Chang

NEW YORK

Published in 2010 by Stewart, Tabori & Chang

An imprint of ABRAMS

Library of Congress Cataloging-in-Publication Data:

Goodall, Jane, 1934–
  Jane Goodall : 50 Years at Gombe / Jane Goodall.
      p. cm.
  Includes bibliographical references and index.
  ISBN 978-1-58479-878-1 (alk. paper)
  1.  Goodall, Jane, 1934–2.  Chimpanzees–Behavior–
Tanzania–Gombe Stream National Park.
3.  Primatologists–England–Biography.
4.  Gombe National Park (Tanzania)  I. Title.

  QL31.G58A3 2010
  599.88509678'28–dc22
      2010004971

Editor: Ann Stratton

Design: Matt Bouloutian, Modern Good

Production Manager: Tina Cameron

The text of this book was composed in Berthold
Akzidenz Grotesk, Adobe Garamond Pro, Craw Modern

Printed and bound in Hong Kong, China

10 9 8 7 6 5 4 3 2 1

Stewart, Tabori & Chang books are available at special
discounts when purchased in quantity for premiums and
promotions as well as fundraising or educational use.
Special editions can also be created to specification.
For details, contact specialsales@abramsbooks.com or
the address below.

115 West 18th Street
New York, NY 10011
www.abramsbooks.com

TO THE MEMORY

OF MY AMAZING MOTHER,

VANNE, WITHOUT

WHOSE WISE GUIDANCE THIS

RESEARCH MIGHT

NEVER HAVE HAPPENED;

TO LOUIS LEAKEY, FOR HIS BELIEF

IN A YOUNG, UNTRAINED WOMAN;

TO RASHIDI KIKWALE, WHO

FIRST INTRODUCED ME

TO THE FORESTS OF GOMBE; TO

DAVID GREYBEARD

AND FLO, WHO INTRODUCED

ME TO THE WORLD OF THE

WILD CHIMPANZEES; AND TO RUSTY,

WHO TAUGHT ME THAT

ANIMALS HAVE PERSONALITIES,

MINDS, AND FEELINGS

LONG BEFORE I MET A

CHIMPANZEE

# A Genuine Heroine

On a rainy street corner in Nairobi, almost fifty years ago, Jane Goodall and I met for the first time. She and I, two young women born in the same year, had no clue then how intertwined our lives would become. Jane had recently begun her chimpanzee project at Gombe under the direction of paleoanthropologist Louis Leakey. Shortly after her astonishing discovery of tool use by the chimpanzees, she was given a small grant by the National Geographic Society to support her fieldwork. I was a *National Geographic* editor on assignment in East Africa, meeting with Leakey and his wife, Mary, to plan photographic coverage of their monumental work at Olduvai Gorge. Before I left our headquarters in Washington, I'd also been told to size up the young blonde woman working with chimpanzees in Tanzania. Perhaps, went the thinking, her project might eventually amount to something of popular interest for *National Geographic*. How accurate that thinking turned out to be. Jane not only became the world's best-known primate scientist, she also became a living symbol for the preservation of our natural world and its animal populations. Her energy and tireless dedication to these very best of causes are legendary.

Over the years, when people learned I worked for the National Geographic Society, a question I could count on being asked was, "Gosh, did you ever meet Jane Goodall?" Well, gosh, I certainly did. I directed the production of her illustrated articles for *National Geographic* and alerted our television and book divisions to take a hard look at this unique scientist. The result? Three *National Geographic* books and four television films. Jane and I became close friends. One day, she asked me to serve on the board of the Jane Goodall Institute, created in 1977. Later I became its president. Long ago I asked Jane why she felt the way she did about animals, why she was adamant we should be kind to them. Her answer has always stayed with me: "We should be kind to animals because it makes better humans of us all."

But let's go back to the beginning. While I was in Kenya in 1962, the Dutch photographer Hugo van Lawick was assigned to work with the Leakeys. I also asked him to look in on Jane's chimpanzee work at Gombe, not really expecting much. Meaningful photographs of wild chimpanzees had always been next to impossible to take, but as the chimpanzees began to adapt to Jane's and then to Hugo's presence in their territory, photographs began to trickle in to my office at *National Geographic*. At first, each time I reviewed another small group of Hugo's transparencies, I shook my head sadly—nothing, absolutely nothing. Then slowly the tide began to turn. Incredible close-ups of heretofore unknown chimpanzee behavior started to appear, photographs that were used to illustrate Jane's early publications and then her best-selling first book, *In the Shadow of Man*. Many of these outstanding shots are in this book you now hold in your hands.

How do you explain Jane's tenacity, her total devotion to the things she's convinced are right? You should have known Vanne Goodall, Jane's amazing, delightful mother. Vanne, who lived in Bournemouth, England, went with Jane to Tanganyika, later renamed Tanzania, in 1960 when the Gombe project began. She and her not-yet-famous daughter endured incredible hardships, but typical of both of them, they stuck it out, and the research camp was established. Vanne eventually returned to England, periodically showing up wherever Jane happened to be in the world, including my office in Washington. "Mary, you're not doing enough to help Jane," Vanne scolded me more than once. "Let's put more articles about her work in your magazine!" Far from resenting her comments, I welcomed them. She was never mean or quarrelsome, just pleasantly tenacious, with an impish sense of humor. Jane obviously inherited her own never-give-up spirit and her sense of humor from her late, much-loved mother.

And Jane has another invaluable quality: She can figure out in an instant any audience she might be facing—the person sitting next to her on an airplane, three people at dinner, a huge filled-to-the-rafters auditorium, or millions of worldwide television viewers. Her talent for communicating her messages is almost uncanny. Some years ago, a women's political action-group in Hollywood set up a small luncheon for Jane with an eye toward helping her raise funds on the West Coast. We numbered about twenty that day at an ultra-chic Santa Monica restaurant—powerful entertainment-industry women, and Jane and me. I'd flown to Los Angeles for the occasion. The luncheon went well, with lots of movie and television chit-chat, lots of fascinating facts and figures. I wondered how Jane would shape her after-dessert talk. She folded her napkin, rose to her feet, placed her hands on the back of her chair, and began a brief, riveting description of her work with chimpanzees, her concern for animals, and the warning that we must care for our planet and everything on it or risk losing it all. She was brief, to the point, and very personal. When Jane sat down, there wasn't a dry eye in the house, and many contributions to the Jane Goodall Institute resulted.

So what's Jane Goodall really like (another question I'm constantly asked)? She's extraordinary, one of the few world-class celebrities who, without question, fully deserves all the respect and adulation she's received. Jane is a superb scientist and a genuine heroine in a world crowded with hero wannabes. I suspect you already know this, but if not you'll surely agree by the time you've finished reading this remarkable book and studying its unique photographs.

## —MARY SMITH

former senior assistant editor at *National Geographic* magazine

## A Message from Jane Goodall

Fifty years ago, in July 1960, I began a study of chimpanzees in the Gombe Stream Chimpanzee Reserve (now Gombe National Park) in the British protectorate of Tanganyika (now Tanzania). I had not attended college then, and it had been difficult for my mentor, the late Louis Leakey, to find money for me. Eventually, though, he got a six-month grant from Leighton Wilkie, a Des Plaines, Illinois, businessman with an interest in human evolution. The British authorities had refused to let a young girl go into the forest alone—so my mother, Vanne, volunteered to accompany me. Bernard Verdcourt, a botanist from the Coryndon Museum where Leakey was curator, offered to drive us there. After some eight hundred miles, over mostly dirt roads, in his overloaded Land Rover, we arrived in Kigoma, a small town on the eastern shores of Lake Tanganyika.

What an arrival. On the other side of the lake, the people of what was then the Belgian Congo (which subsequently became Zaire and today is the Democratic Republic of the Congo) had risen up against the white settlers. Kigoma was overflowing with refugees, most of them sleeping on mattresses on the floor of a large Belgian warehouse at the port. The first night we three travelers shared one small room in the only hotel, but a Belgian family was desperate, and so we moved out and set up our tents, as directed, in the grounds of the prison. (It was well guarded and would be safe, we were told.) And we helped the citizens of Kigoma to feed the refugees, making hundreds and hundreds of Spam sandwiches.

It was two weeks before we could embark upon the last phase of the journey, for it was feared that the Africans in the Kigoma region might follow the example of the Congolese. But this did not happen, and so, on July 14, Vanne and I set off in the government launch, the *Kibisi*, on the twelve-mile journey to the Gombe Stream Game Reserve. The tiny aluminum boat that would be our only link with civilization was on board, along with provisions for several weeks, and Dominic Charles, who had been chosen to look after the camp and cook our food. Bernard had returned to Nairobi, convinced (he later confessed) that he would never see us again.

On arrival we were greeted by the two resident game scouts and Iddi Matata, who, we discovered later, was the most infamous witch doctor in the region. Fortunately he decided to extend his protection over the two strange white women. The cumbersome ex-army tent that Vanne and I would share was pitched near a small stream in a little clearing close to the lakeshore. Then, with about an hour of daylight left, I set off to explore the forested slope rising above the camp.

If I close my eyes, I can recall my utter joy as I sat in a clearing looking down at the lake shimmering in the evening sun. A troop of baboons paused to threaten me, barking at the intruder. There was the smell of a recent bush fire and the gentle falling notes of mourning doves. After supper around the campfire, I pulled my camp bed outside and lay looking up at the stars wondering if any of this could really be happening. Had my childhood dream actually come true?

During the months that followed, I sometimes despaired that our money would run out before the chimpanzees lost their fear of the strange white ape who had invaded their forest world—and that would be the end. But Vanne was always pointing out how much I was actually learning—how the chimpanzees made nests each night, traveled in groups of different sizes, ate fruits and leaves and flowers, and so on.

Vanne had to leave after five months—about a week before some twenty young men from Mwamgongo village, located to the north of Gombe, invaded my camp. They wanted to drive me away so that they could move into the reserve to cultivate the fertile valleys, and they were armed with *pangas* (machetes). Expecting to find me in my tent, they arrived at six o'clock in the morning—but, as always, I was already up the mountain. Thank goodness Vanne was not there, either. They cut down trees all along the stream and then left. I still remember the sick feeling I had when I got back and saw the devastation. Rashidi Kikwale, who had been appointed to accompany me into the field for the first few months, reported the incident to the authorities in town, and they came to fetch me in the *Kibisi*. I was forced to stay in Kigoma until the ringleaders, identified by Rashidi (who himself came from Mwamgongo), had been arrested and put in jail.

Then I returned to the forest, and soon after I saw David Greybeard, the first chimp to lose his fear of me, using grass stems as tools to fish for termites and even stripping leaves from a twig to *make* a tool. It was this breakthrough, the startling discovery that we humans were not the only beings to use and make tools, that prompted the National Geographic Society to make the first of many grants, enabling me to continue my research. Meanwhile, Tanganyika, which had been part of German East Africa until Word War II, made a peaceful transition to independence in 1961, under President Julius Nyerere. In 1964 it joined with Zanzibar to form the Republic of Tanzania.

In 1962, the National Geographic Society sent Hugo van Lawick to photograph and film the chimpanzees. This resulted in an article in their magazine and the documentary film *Miss Goodall and the Wild Chimpanzees,* which took the Gombe chimpanzees into the living rooms—and hearts—of people around the world. Two years later, Hugo and I married and our son, Hugo Eric Louis, nicknamed Grub, was born in 1967. By that time I had obtained my Ph.D. degree from the University of Cambridge in England. The Gombe Stream Research Centre was established, affiliated with the University of Dar es Salaam in Tanzania, the University of Cambridge, and Stanford University in the United States, and many students from Europe and America, along with our well-trained Tanzanian field staff, were learning ever more about Gombe's chimpanzees, baboons, red colobus monkeys, and so on. A major grant for this work was obtained from the William T. Grant Foundation. And the game reserve was gazetted by parliament as Gombe National Park, giving the area greater protection.

In the 1970s, the Ujamaa movement, initiated by the socialist government of Julius Nyerere, forced scattered communities to relocate into villages and practice communal farming. This resulted in rapid deforestation of the hills around Gombe as villagers tried to create new fields for their crops. At the same time, some of the refugees fleeing the ethnic violence that repeatedly broke out in Burundi (the border is not far from the northern park boundary) sought refuge in the hills around Gombe. And then civil war broke out on the other side of the lake in what was then Zaire. In May 1975, four of the students from Gombe were kidnapped by the rebel forces of Laurent Kabila and held for ransom. The money was paid, and they were eventually returned, but the incident led to the temporary closing of the research station to foreigners. All traffic up and down the lake was carefully monitored and a strict curfew enforced.

By this time, Hugo and I had parted ways, and I had married Derek Bryceson, then director of national parks in Tanzania. He made it possible for me to visit Gombe during that time and also helped me maintain the team of Tanzanian field staff, dedicated men who were able to continue regular observations. But because there was no Ph.D. scientist in residence at Gombe, the William T. Grant Foundation withdrew its support. The financial instability that followed prompted Ranieri and Genevieve (Genie) di San Faustino to set up the Jane Goodall Institute (JGI) as a vehicle for fund-raising—initially solely to maintain the research at Gombe.

In 1986, during a conference in Chicago, researchers from across Africa presented information about the rapid decline of chimpanzees throughout their range countries due to human population growth, habitat destruction, and hunting. Chimpanzee infants had been captured in large numbers (by killing the mothers) initially for the international wild-animal trade (for medical research and entertainment) and subsequently as part of the burgeoning bushmeat trade—the commercial hunting of wild animals for food. At the conference there was also a session about the treatment of chimpanzees in medical-research laboratories and other captive situations, such as the circus, movies, pets, and so on. How could I continue my idyllic life—collecting data in the forest, writing papers, teaching at Stanford twice a year? I felt compelled to do what I could for the chimpanzees, and so I took to the road to raise awareness around the world about their plight.

Meanwhile, the situation in eastern Zaire remained unstable; refugees were continually arriving, some of them hiding in the hills around Gombe. In the early 1990s, I flew over the area in a small plane and was horrified to see the degradation of the land outside the tiny national park. How could we even try to protect the chimpanzees when people were struggling to survive? And so in 1994, JGI initiated the TACARE program, a community-based conservation program that has improved the lives of the villagers living around Gombe. Today, these villagers have become our partners in conservation. Trees are springing up on the devastated slopes, and through our Roots & Shoots programs in schools, children are learning about the chimpanzees and the need to protect them.

The Gombe Stream Research Centre has gradually grown again and, as you will read in the following pages, exciting new studies are ongoing. I get back there twice a year, though these visits are short. Yet, I can climb up to the peak from where I first observed the chimpanzees at a distance, or sit by the waterfall in Kakombe Valley, and recapture the wonder of the early days, renewing my energy, and absorbing the peace of the forest to sustain me for the tough months on the road. And although, along with Rashidi, Derek, Hugo, and my mother, my old chimpanzee friends have all passed away, their children and grandchildren are now roaming the forests.

One thing is certain: As the years go by, we shall continue to learn new things about these closest relatives of ours. And there will be more people—in Africa and around the world—who will join us in the fight to protect them and their forest homes.

Jane Goodall

Glitta reaches for flying termites in the treetops at Gombe's Peak Ridge.

# THE BEGINNING

YAHAYA ALMASI BOWS HIS HEAD; HIS WEATHERED FACE IS WRINKLED UP IN DEEP CONCENTRATION. IN A SOFT VOICE, HE SPEAKS ABOUT THE FIRST TIME HE HEARD OF JANE GOODALL, THE STRANGE YOUNG *MZUNGU* (WHITE PERSON) WHO CAME IN 1960 TO THE FOREST NEAR HIS VILLAGE OF BUBANGO, TANZANIA, TO STUDY CHIMPANZEES.

IT'S THE CHIMPANZEES DANCING IN THE FOREST, SINGING,

# "WE USED TO BE PEOPLE, BUT NOW WE ARE NOT."

Y AHAYA HAD HEARD OF THE CHIMPANZEES THROUGHOUT HIS LIFE, AND HE'D HEARD THEIR WILD CALLS. HE'D HEARD OF THEIR HUMANLIKE BEHAVIORS, THEIR GREAT STRENGTH, AND THEIR COMMUNITIES IN THE DEEP RECESSES OF THE NEARBY FOREST. BUT WHAT HE REMEMBERED MOST WAS HIS GRANDMOTHER'S TALE OF THEIR ORIGIN.

On the occasion of the Darkness Twice, his grandmother had said, when the moon's shadow covers the sun in the heat of the day, everyone must hide in their houses, secure with food and firewood. If you're caught outside when the darkness descends, you'll become a chimpanzee—just like those in the forest. Listen carefully, she continued, when you hear the distant pounding of the tree buttress. It's the chimpanzees dancing in the forest, singing, "We used to be people, but now we are not."

So when Yahaya heard of Jane Goodall's plans to study the chimpanzees, he thought she surely must be a brave woman—perhaps coming armed with voodoo—to want to live among these strange animals of the rainforest. Later he would learn that the village elders were not alone in their belief that humans and chimpanzees share a common origin. Dr. Louis Leakey, the famed paleontologist and anthropologist, believed wild chimpanzees could provide a glimpse into the lives of early humans. By studying the chimpanzees' behavior, he thought—their food gathering, their daily habits, and their social relations—we might have a better understanding of the evolution of man.

When Jane began her study, chimpanzee populations could be found throughout the equatorial belt and in the forests of West Africa. At that time, before human encroachment and the onslaught of commercial logging, most of this habitat was

ABOVE: Echo cradles her baby Emela in Gombe National Park.

TOP: Gaia in Gombe National Park

OPPOSITE: Mist forms over the verdant hills of the Kakombe Valley, Gombe National Park.

PREVIOUS PAGES: Fifi studies Jane.

# THE CHICKEN COOP

**JANE'S MOTHER, VANNE—A REMARKABLE WOMAN IN HER OWN RIGHT—WATCHED HER ELDER DAUGHTER'S CURIOSITY ABOUT THE ANIMAL WORLD DEVELOP FROM A VERY EARLY AGE.**

On one occasion, Vanne returned home during the war years to find the house curiously empty; everyone, she soon learned, was out searching for Jane, who at this point had been missing for several hours.

By seven o'clock that evening, expressions were turning grave. "I don't remember who saw her first—a small, disheveled figure coming a little wearily over the tussocky field by the hen houses," Vanne wrote. "There were bits of straw in her hair and on her clothes, but her eyes, dark ringed with fatigue, were shining. 'She's found,' someone called out and soon the searchers were all back and gathered around Jane in the stable yard. 'Wherever have you been?'

'With a hen.'

'But you've been away for nearly five hours. What can you possibly have been doing with a hen all that time?'

'Well, you see, I had to find out how hens lay eggs, so I went into a hen coop to find out, but as soon as I went in, the hens went out, so I went into an empty coop and sat in the corner and waited until a hen came in who didn't mind me there.'

'So, then what happened?'

'A hen came at last.' Jane's eyes glowed with inner radiance. 'It was a long time, but she came at last and then she laid an egg. I saw her. So now I know how a hen lays an egg.'

Dusk was fast gathering under the trees as we made our way back to the house. I had Jane firmly by the hand. She had just spent five hours crouched double in a stuffy hen coop, but the result had made it all worthwhile. She had successfully, and to her own entire satisfaction, completed her first animal-research program. She had observed a hen lay an egg." (From the book *Jane Goodall by Her Mother*, Vanne Goodall.)

OPPOSITE, LEFT: Young Jane with her mother, Vanne

OPPOSITE, TOP: Jane with her toy chimpanzee, Jubilee, a gift from her father. The toy commemorated the birth of the London Zoo's first captive-born chimp. Jane loved the toy and carried it with her everywhere for many years.

OPPOSITE, BOTTOM: Jane's love of animals was evident from an early age.

quite remote. Leakey chose the chimpanzee habitat of Gombe Stream Chimpanzee Reserve, on the shore of Lake Tanganyika in the northwest corner of Tanzania, because it was easily accessible from the lakeshore. At that time, the reserve was surrounded by forest on three sides and a lake on the other. It comprised thickly forested valleys giving way to open woodland and bare ridges as it rose up from the lakeshore. It offered, Leakey believed, an ideal location for studying the behavior of wild chimpanzees. His next step was to find someone to conduct the study. Fortunately for Leakey and for the chimpanzees, he found Jane Goodall.

Jane was born in London, England, in 1934, the first daughter of Mortimer and Vanne (short for the Welsh name *Myfanwe*, pronounced "Van") Goodall. Throughout her childhood, Jane showed a fascination with animals. When she was but eighteen months old, her nanny ran from Jane's room to tell Vanne that Jane had a handful of "horrible, pink, wriggling worms" in her bed. "They're under her

pillow," the nanny said, "and she won't let them go." Vanne never forgot that day. "A peach-colored light from the setting sun was flooding the nursery," Vanne later recounted. "Jane's eyes were already closing, one hand was out of sight beneath the pillow. I pointed out that the little creatures would find it altogether too hot and stuffy beneath the feathers. And after a tear or two, she agreed to come with me to the dusky garden and return them to their rightful home. We did not realize then that the incident meant any more to Jane than the sorrow normally suffered by children when they lose a pet. But on looking back, I think it did. I believe her absorbing interest in the animal world was, even then, oddly objective. She was curious about them, and this insatiable curiosity about life, its origins and complexities, its mysteries and failures, has never left her."

From earthworms, Jane turned her attention to hens, dogs, and other animals she met around her own garden. When she began to read, the stories of Doctor Dolittle and Tarzan unlocked her imagination. Before long, she dreamed of going to Africa.

BELOW: Jane with paleontologist Louis
Leakey, soon after beginning her study of
the chimpanzees

OPPOSITE, LEFT: Young Jane in her garden

OPPOSITE, RIGHT: Jane and Vanne, in their tent
at Gombe, preserving specimens of plants.

BELOW: Jane with paleontologist Louis
Leakey, soon after beginning her study of
the chimpanzees

Vanne's friends cautioned her about Jane's ambitions. "Tell Jane to dream about something she can achieve," they would say. "Don't give her false hope." But Vanne felt differently, and when Jane spoke of one day living in Africa to study animals, Vanne told her, "If you really want something, and you work hard enough, take advantage of opportunities, and never give up, you will find a way."

Jane's opportunity came when a school friend whose family had moved to Kenya invited her to visit. Jane saved up her boat fare working as a waitress and, shortly after her arrival in Nairobi, made an appointment to meet Leakey, then curator of the Coryndon Museum.

After one year of Jane working as Leakey's secretary in the museum, assisting him and his wife, Mary, with a paleontological dig in the now-famous Olduvai Gorge, Leakey asked Jane to study the chimpanzees of Gombe Stream Chimpanzee Reserve. She had no scientific training; indeed, she did not even hold an undergraduate degree. But Leakey was impressed with Jane's unending knowledge of animals and their behavior. He saw that she had an insatiable curiosity about the animal world, a strong determination to find the answers, and the necessary patience to await their discovery. Jane also had physical endurance and could go for long stretches without needing to eat or rest, a quality she inherited from her father.

Yet there were a few obstacles to overcome before Leakey could send Jane out. She needed the funds to support her preliminary investigation—funds for travel, food, and guides into the forest. With Leakey's help, she received a small grant from the Illinois-based Wilkie Foundation, which supported studies of primates. They then faced the concern of the British authorities in what was then the protectorate of Tanganyika, who refused to allow a young European woman to work in the forests of the "Dark Continent" by herself. But Leakey prevailed. If they wouldn't let Jane live there alone, surely it would be fine if she had a companion. And Vanne, who had always supported and encouraged Jane, volunteered to fill that position.

Nineteen months later, on July 14, 1960, Jane and Vanne arrived at Kasekela, midway along the ten-mile shoreline of Gombe Stream Chimpanzee Reserve. The site was chosen because two African game scouts had huts nearby. Jane and her mother were accompanied by Dominic, their African cook.

The early days were a struggle. Jane rose before dawn to search for the elusive chimpanzees. She trudged along the clear streams, hiked up the steep slopes, and crawled through the dense undergrowth to follow a far-off call. When she occasionally spotted a group of chimpanzees gathered in a tree, they ran away before she could get close enough to watch. But Jane was patient. She sat hour after hour on "the peak," watching the chimps through binoculars.

Slowly, the chimpanzees began to accept the presence of this strange white ape. And as she began to recognize and distinguish each unique chimpanzee face and character, Jane gave the chimpanzees names. (Later she would be told that it would

"IF YOU REALLY WANT SOMETHING, AND
YOU WORK HARD ENOUGH, TAKE ADVANTAGE
OF OPPORTUNITIES, AND NEVER GIVE UP, YOU WILL FIND A WAY."

LEFT: Jane at Leakey's paleontogical dig at Olduvai Gorge

BELOW: Jane and Flo's infant Flint greet each other.

OPPOSITE: Vanne set up a clinic where she provided basic medical treatment to local villagers.

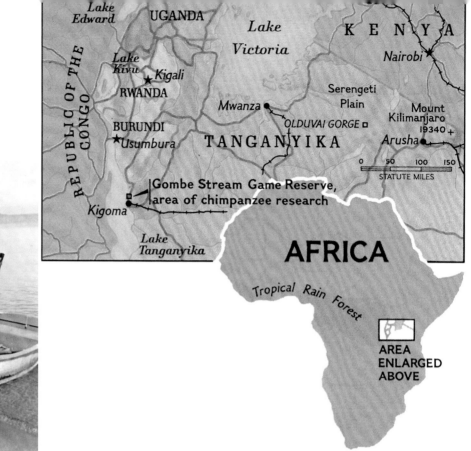

ABOVE: Vanne with the expedition boat driver, Hassan, on the shore of Lake Tanganyika

OPPOSITE: Jane searches with binoculars for a glimpse of chimpanzees.

have been more scientific to give them numbers.) Soon she was documenting the lives of Flo, her daughter Fifi, David Greybeard, Mr. Worzle, Mike, Goliath, and all the rest.

With the Wilkie Foundation funding soon to run out, Jane was desperate for a discovery significant enough to continue the funding. The discovery she made in November 1960 not only continued her financing but also changed the way humans view the animal world. Jane saw David Greybeard strip the leaves off a twig before poking it into a termite mound to retrieve the tasty insects. David was modifying a natural object for a specific purpose. He was not only using but also making a tool.

When the Wilkie grant money was completely gone, the National Geographic Society (NGS) stepped in. On March 13, 1961, the NGS Committee for Research and Exploration issued Jane Goodall a grant of $1,400. It was the first of twenty-seven grants from the Society for a study that has gone on for fifty years—one of the longest continuous field studies ever conducted on a wild-animal group.

Although Vanne returned to England after five months, her time at Gombe had laid the groundwork for Jane's relationship with the local people. While Jane trekked through the forest from dawn to dusk to learn about chimpanzees, Vanne had stayed at the camp dispensing aspirin and other simple first-aid remedies to local villagers and fishermen.

One such fisherman was Yahaya Almasi. When he came into the small makeshift "clinic" (four posts with a thatched roof) to receive medicine for malaria, he met the mother of the woman who studied the chimpanzees. He and his fellow fishermen sometimes discussed Jane and the chimpanzees as they fished on the lake at night, and Yahaya remembered his grandmother's tales of how chimpanzees came to be. His curiosity grew.

In 1978, Yahaya left his life as a fisherman to join Jane and some men from the surrounding villages in the groundbreaking research of Tanzania's chimpanzees. Soon after, he stopped believing his grandmother's tale that chimpanzees were once people, but he joined the rest of the world in marveling at the amazing similarities between humans and chimpanzees. And this, in the end, is the insightful gift Jane Goodall's determination has given to the world.

Jane's first expedition to Gombe was sparingly equipped—there was very little funding. Even so, the Land Rover that transported Jane, her mother, and their belongings from Nairobi to Kigoma on the shores of Lake Tanganyika was overloaded. The trip took four days, with botanist Bernard Verdcourt driving. Food, pots and pans, and kerosene (for lamps) were purchased in Kigoma. Eventually everything was transported to Gombe by boat.

# THE CHIMPANZEES

"HOW CAN SHE POSSIBLY BE SO UGLY?" THOSE WERE THE WORDS OF AN EARLY VISITOR TO GOMBE UPON SEEING OLD FLO, ONE OF THE RESERVE'S MOST BELOVED CHIMPANZEES. INDEED, FLO WAS NOT A BEAUTIFUL CHIMPANZEE—AT LEAST TO HUMAN EYES. SHE HAD RAGGED EARS, A BULBOUS NOSE, AND TEETH WORN DOWN TO THE GUMS. BUT SHE WAS ONE OF THE MOST POPULAR FEMALES AS FAR AS THE MALE CHIMPANZEES WERE CONCERNED, AND PEOPLE ALL OVER THE WORLD ADMIRED HER CHARACTER AND HER MOTHERING SKILLS, AS A RESULT OF THE STORIES TOLD BY DR. GOODALL IN HER BOOKS AND ARTICLES.

WE HAVE FOLLOWED THE LIVES, LOVES, AND HARDSHIPS
# OF THE CHIMPANZEES OF GOMBE
# AS IF THEY WERE CHARACTERS
# IN A SOAP OPERA.

# J

ANE REVEALED TO THE WORLD THAT EACH CHIMPANZEE IS AN INDIVIDUAL WITH HIS OR HER OWN PERSONALITY, CHARACTERISTICS, AND QUIRKS. INDEED, SINCE JANE'S FIRST *NATIONAL GEOGRAPHIC* ARTICLE LANDED IN MAILBOXES IN AUGUST 1963, WE HAVE FOLLOWED THE LIVES, LOVES, AND HARDSHIPS OF THE CHIMPANZEES OF GOMBE AS IF THEY WERE CHARACTERS IN A SOAP OPERA. THEIR BEHAVIORS AND LIFE STORIES NEVER FAIL TO INTRIGUE AND INSPIRE.

One of the first chimpanzees Jane came to know was David Greybeard. "David was less afraid of me from the start," she wrote in her book *In the Shadow of Man*. "I was always pleased when I picked out his handsome face and well-marked silvery beard in a chimpanzee group, for with David to calm the others, I had a better chance of approaching to observe them more closely."

David Greybeard was the first to come into Jane's camp, to feast on the ripe fruit of the oil nut palm that grew there, and one day he "stole" bananas from Jane's tent. When the palm stopped fruiting, Jane left bananas around the camp for David, should he wander through. He soon began to bring others with him— Goliath, William, and sometimes a younger male, Faben—helping them to see that Jane was not a threat.

It was David who provided Jane with her two first, and most significant, discoveries. For not only was he the first to demonstrate that chimpanzees use and make tools, but it was he whom she saw eating the remains of a bushpig infant. Before this, chimpanzees were thought to be primarily vegetarians and fruit eaters.

And then came a truly magical experience.

"One day, as I sat near him at the bank of a tiny trickle of crystal-clear water,

ABOVE: Two of Jane's favorite chimps, David Greybeard (left) and young Fifi

TOP: Jane offers a banana to David Greybeard. In later years, Jane and the other Gombe researchers stopped feeding and touching the chimps.

OPPOSITE: Wildlife photographer Hugo van Lawick was hired in 1962 by the National Geographic Society to photograph Jane's work at Gombe. The two, who were married in 1964, collaborated for *National Geographic*, spreading the Gombe story around the world.

PREVIOUS PAGES: Flo uses a long piece of straw as a tool to fish termites from their nest.

I saw a ripe red palm nut lying on the ground," Jane wrote in her book *In the Shadow of Man.* "I picked it up and held it out to him on my open palm. He turned his head away. When I moved my hand closer he looked at it, and then at me, and then he took the fruit, and at the same time held my hand firmly and gently with his own. As I sat motionless he released my hand, looked down at the nut, and dropped it to the ground."

"My own relationship with David was unique—and will never be repeated," Jane wrote in her 1986 book *The Chimpanzees of Gombe: Patterns of Behavior.* "When David disappeared during an epidemic of pneumonia in 1968, I mourned for him as I have for no other chimpanzee before or since."

Old Flo was another early visitor to Jane's camp. She was accompanied by two of her offspring, juvenile Figan and infant Fifi. Then, adolescent Faben began to arrive with them, suggesting by his behavior that he was part of the family. This gave Jane the first glimpse of the close and lasting bonds that develop between mothers and their older offspring. When Flo became sexually receptive, she was followed into the camp by many male suitors, each hoping for a chance to mate. When the males discovered bananas in the camp, they, too, became regular visitors.

It was Flo's personality that endeared her to Jane. "She was aggressive, tough as nails, and easily the most dominant of all the females at that time," Jane wrote in her book *In the Shadow of Man.* She was a superb mother—easygoing, tolerant, playful, and protective, but with enough discipline to keep Fifi in line—important for the child's development. In 1964, Flo gave birth to another infant, whom Jane named Flint. Jane delighted in watching as Flo cared for her new son. And she noted how big sister Fifi was absolutely fascinated by the baby and desperate to have contact with her new brother.

ABOVE: Flo with her playful infant Flint

OPPOSITE: David Greybeard has been given a handful of bananas.

Flo, inevitably, began to show her age. She wasn't strong enough to wean Flint when the time came, nor to care for her new infant Flame, born when Flint was only four and a half years old (the normal birth interval is five years or more). Flame disappeared during a time when Flo was too ill even to climb into a nest. Though she recovered somewhat after Flame's death, she did not have the strength to discipline Flint, who insisted on riding on her back and sleeping in her nest. When Flo died in 1972, Flint, abnormally dependent on his old mother, fell into a state of depression. Lethargic and gaunt, he was not interested in eating. Jane poignantly described Flint's last days in *Through a Window*: "The last journey he made, pausing to rest every few feet, was to the very place where Flo's body had lain. There he stayed for several hours, sometimes staring and staring into the water. He struggled on a little further, then curled up—and never moved again."

For Jane, Flo's death left a void at Gombe—the two had spent so many hours together. (Flo is the only chimpanzee honored with an obituary in Britain's *Sunday Times*.) But her legacy lives on in her descendants—members of the highly success-ful "F" family. Flo's son Figan became alpha and her grandsons Freud, Frodo, and Ferdinand all achieved alpha status as well. In addition, her two granddaughters, Fanni and Flossi, are high-ranking mothers, each with several offspring of their own.

ABOVE: Flo, Flint, and Fifi rest in a nest thirty feet above the ground.

OPPOSITE: Jane often took to the trees to see above the fourteen-feet-tall grasses.

# Gombe's F FAMILY family tree

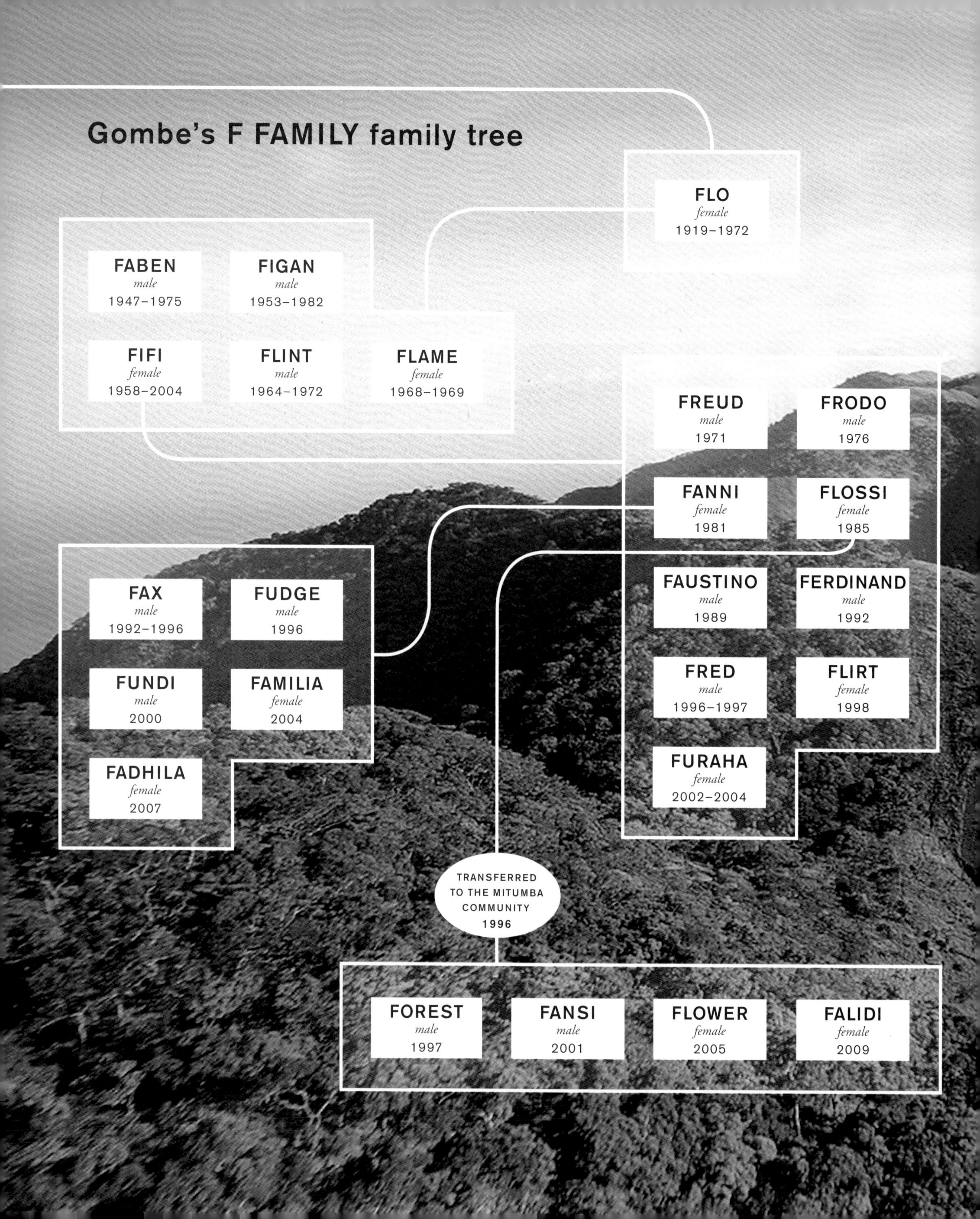

**FLO**
*female*
1919–1972

**FABEN**
*male*
1947–1975

**FIGAN**
*male*
1953–1982

**FIFI**
*female*
1958–2004

**FLINT**
*male*
1964–1972

**FLAME**
*female*
1968–1969

**FREUD**
*male*
1971

**FRODO**
*male*
1976

**FANNI**
*female*
1981

**FLOSSI**
*female*
1985

**FAX**
*male*
1992–1996

**FUDGE**
*male*
1996

**FAUSTINO**
*male*
1989

**FERDINAND**
*male*
1992

**FUNDI**
*male*
2000

**FAMILIA**
*female*
2004

**FRED**
*male*
1996–1997

**FLIRT**
*female*
1998

**FADHILA**
*female*
2007

**FURAHA**
*female*
2002–2004

TRANSFERRED
TO THE MITUMBA
COMMUNITY
1996

**FOREST**
*male*
1997

**FANSI**
*male*
2001

**FLOWER**
*female*
2005

**FALIDI**
*female*
2009

# Gombe's G FAMILY family tree

**MELISSA**
*female*
1949–1986

**GOBLIN**
*male*
1964–2004

**GREMLIN**
*female*
1970

**GETTY**
*male*
1982–1986

**INFANT**
1987–1987

**GENIE**
*female*
1976–1976

**GIMBLE**
*male, twin*
1977–2007

**GALAHAD**
*male*
1988–2000

**GAIA**
*female*
1993

**GYRE**
*male, twin*
1977–1978

**GROUCHO**
*male*
1985–1986

**GOLDEN**
*female, twin*
1998

**GLITTA**
*female, twin*
1998

**GIMLI**
*male*
2004

**GIZMO**
*male*
2009

**GODOT**
*male*
2006–2006

**INFANT**
*twin*
2008–2008

**INFANT**
*twin*
2008–2008

**GHAFLA**
*male*
2009

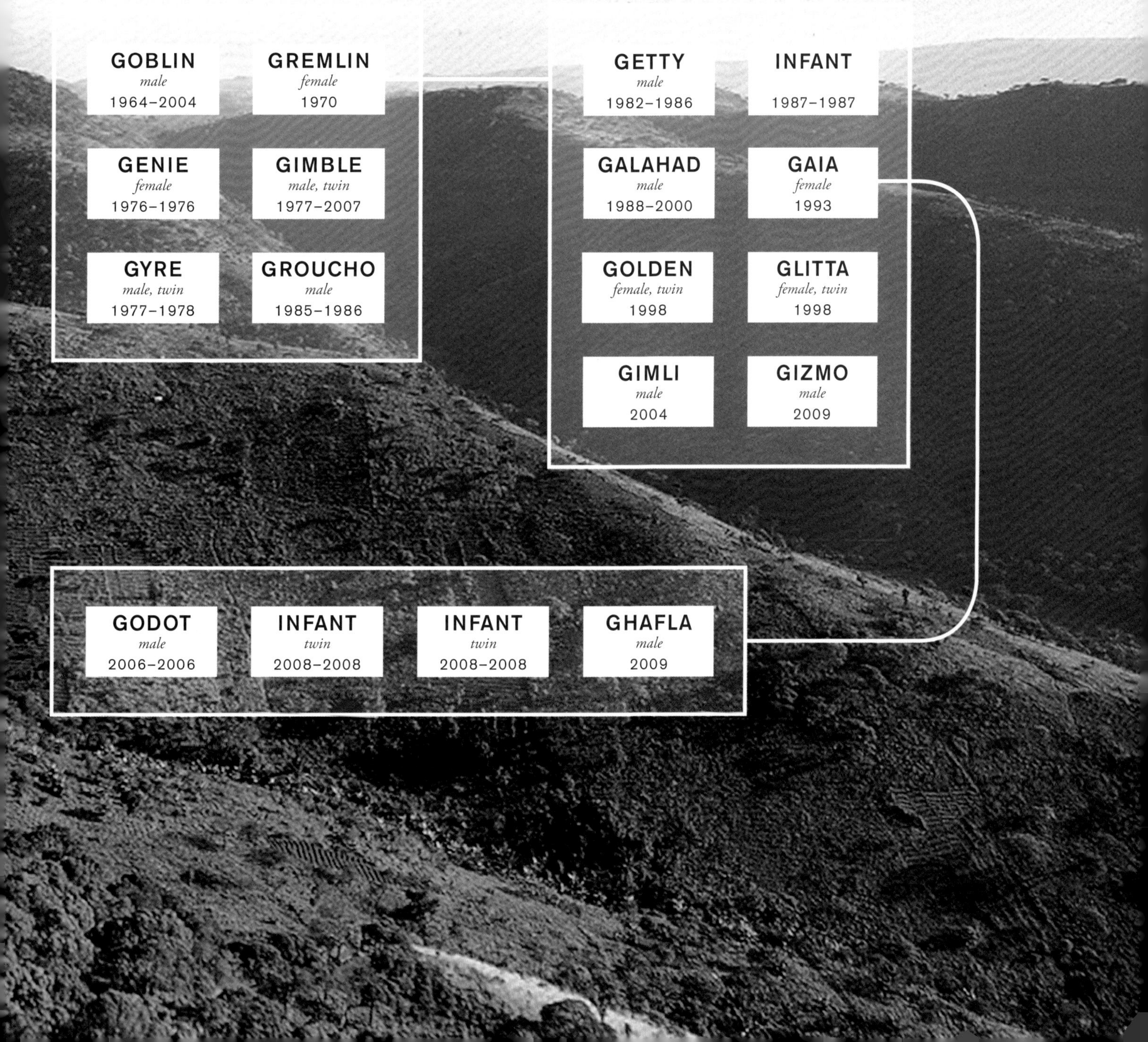

LEFT: Each evening in her tent, Jane would write up data from her field notebooks, recounting the chimp behavior she observed that day. During her first year at Gombe, she wrote by hand on lined three-ring notebook paper. Later she managed to get a portable typewriter.

BELOW: Jane sips coffee and watches the sun set over Lake Tanganyika.

FOLLOWING PAGES: In her early days at Gombe, Jane spent many hours sitting on a high peak with binoculars or a telescope, searching the forest below for chimpanzees.

# TRIBUTE TO FIFI

## FLO'S DAUGHTER FIFI WAS A HIGHLY CURIOUS AND FEARLESS YOUNGSTER, THRIVING UNDER FLO'S SKILLFUL MOTHERING AND THE PROTECTION OF HER TWO OLDER BROTHERS, FIGAN AND FABEN.

Fifi grew to be a top-ranking female, a central player in Gombe's Kasekela group and one who contributed greatly to our understanding of mothering, sexual behavior, the female dominance hierarchy, and much more. She was extremely successful reproductively, giving birth to her ninth baby at the age of forty-four. (Most females don't raise more than two or three offspring to reproductive maturity.) Despite the typical high rate of infant mortality, Fifi only lost two of hers.

Jane and Fifi had a strong and somewhat mysterious connection, even after Jane stopped daily field research in 1986. Whenever Jane managed to get to Gombe for a few days (twice a year) to "recharge her batteries," Fifi inevitably appeared as though to greet her, often on the second day. She and Jane would sit quietly, as they had done many times before. It was like two old friends enjoying a happy reunion, seeming to commune in silence.

Fifi's disappearance in 2004 was a shock and a profound loss for researchers at the Gombe Stream Research Centre and especially sad for Jane. No one is sure what happened—Fifi simply went missing, along with her last infant. It was feared that she might have been attacked by a neighboring community.

RIGHT: Curious Fifi spoils Hugo's shot when she sees her reflection in his camera lens.

OPPOSITE: Both studying each other, Jane watches Fifi while drinking coffee in her tent, and Fifi perches on the guy-rope, staring back at Jane.

## A Research Center is Born

In 1962, Jane temporarily left Gombe to attend the University of Cambridge and pursue a doctorate in ethology (the study of animal behavior). She was only the eighth person in the history of Cambridge to be allowed to work for a Ph.D. without first having received a bachelor's degree. To avoid losing valuable data during her absence, she set up a program for research assistants to continue following and recording the behavior of the chimpanzees while she was away. And so began the Gombe Stream Research Centre.

Since its inception in 1965, the Gombe Stream Research Centre has become a launching pad for hundreds of students who long to follow in Jane Goodall's footsteps, many of whom are now respected primatologists in their own right and many of whom retain close links with Gombe and its chimpanzees. Jane also opened the door for women, who have played a leading role in the study of chimpanzees and other great apes.

A mother and her infant rest in the treetops in Gombe.

ABOVE: Current residents of Gombe—(left to right) Bahati, Baroza, and Sheldon, a former alpha male

OPPOSITE: Kris, former alpha male

FOLLOWING PAGES: Jane and Emmanuel Mtiti, head of the Greater Gombe Ecosystem Program, view the regenerating forest outside Gombe National Park.

# WHAT WE HAVE LEARNED

THE SCIENTIFIC RESEARCH DR. GOODALL BEGAN IN 1960 HAS FOCUSED ON THE

ROUGHLY ONE HUNDRED CHIMPANZEES LIVING IN THE PARK. IT ALSO INCLUDES

A LONG-TERM STUDY OF THE BEHAVIOR OF BABOONS, AS WELL AS SHORT-TERM

STUDIES OF RED COLOBUS, RED TAIL, AND BLUE MONKEYS.

A HUGE BODY OF WORK HAS COME OUT OF GOMBE FROM JANE AND OTHER RESEARCHERS—MORE THAN TWO HUNDRED SCIENTIFIC PAPERS, THIRTY-FIVE DOCTORAL THESES, THIRTY BOOKS, MANY FILMS, HUNDREDS OF ARTICLES AND SECONDARY WRITINGS, AND HUNDREDS OF LECTURE TOURS AND CONFERENCES. BUT WHAT HAS THE DILIGENT WORK OF GOMBE'S RESEARCHERS TAUGHT US?

ABOVE: Goblin seeks support from Frodo during a tense moment with the other males.

BELOW: Infant chimpanzees spend hours every day playing with other youngsters.

OPPOSITE: Fanni cradles her infant Fax.

PREVIOUS PAGES: Fifi fishes termites from a nest.

## Biology and Behavior

The structure of the chimpanzee brain is startlingly similar to that of the human brain. Many aspects of chimpanzee behavior and social relations, emotional expression, and needs are also similar to those of humans. Various cognitive abilities once regarded as unique to humans have been convincingly demonstrated in chimpanzees, including reasoned thought, abstraction, generalization, symbolic representation, and the concept of self. Nonverbal communication includes hugs, kisses, pats on the back, play tickling, swaggering, punching, and so on. Chimpanzees also express many of the same emotions, such as joy, sadness, fear, and despair.

The chimpanzee developmental cycle is not very different from that of humans. Infancy lasts for five years, followed by childhood, and then adolescence, which lasts until age thirteen. The Gombe chimps start to look old when they are about forty-five years of age. In captivity, they can live for sixty-five years or more.

ABOVE: Frodo pant-hoots from his day nest.

OPPOSITE, LEFT: Chimpanzees communicate in some of the same ways that humans do—through touch, posture, and gesture, as well as sounds. They also show affection with hugs and kisses.

OPPOSITE, TOP RIGHT: For about the first five months, an infant clings to its mother's breast and belly. After that, an infant rides on its mother's back until it is about four years old. This mother is transporting two offspring.

OPPOSITE, BOTTOM RIGHT: One of Jane's earliest discoveries was that chimpanzees are not strictly vegetarian. They hunt small and medium-size mammals, and after a kill, they share the carcass. Here, Fifi is seen dominating the adult males in her group to keep this prize, a young bushbuck. She did not share until completely satisfied.

FOLLOWING PAGES: Current Gombe residents

There are close parallels between the chimpanzee infant and the human child. Both have the capacity for endless romping and play, are highly curious, learn by observation and imitation, and, above all, need constant reassurance and attention. For both, affectionate physical contact is essential for healthy development.

Within their home range, chimpanzees have no special sleeping sites and the distance they travel each day depends on the abundance of food. Each chimpanzee builds a new nest every night—if he does use an old nest, he bends fresh branches over it. Infants will sleep with their mothers until the age of five or until the next baby is born. In the dry season, the chimpanzees usually take a midday siesta on the forest floor in the shade of the trees, but during the rainy season, they often build nests in the trees for daytime naps, thus avoiding the cold, damp ground. Interestingly, they make no serious attempt to shelter themselves from the rain.

In her *National Geographic* article "My Life Among Wild Chimpanzees," Jane reported: "The construction of a nest, I found, is simple and takes only a couple of minutes. After choosing a suitable foundation, such as a horizontal fork with several branches growing out, the chimpanzee stands on this and bends down a number of branches from each side so that the leafy ends rest across the foundation. He [or she] holds them in place with his feet. Finally he bends in all the little leafy twigs that project around the nest, and the bed is ready. But the chimpanzee likes his comfort, and often, after lying down for a moment, he sits up and reaches out for a handful of leafy twigs, which he pops under his head or some other part of his body. Then he settles down again with obvious satisfaction."

The Gombe chimpanzees eat fruit, leaves, stems, seeds, and flowers. But they will also feast on ants, termites, caterpillars, birds' eggs, and honey. Occasionally they hunt, kill, and eat small- and medium-sized mammals, especially young bushpigs, monkeys, or antelope. If a younger chimp has made the kill, the carcass is often divided among a small group. But if the prey belongs to a high-ranking male, other chimpanzees gather around and beg for shares, which they may or may not get.

## Tool Making

Before Jane began her research, it was thought that the ability to craft and use tools was specific to *Homo sapiens*. We now know that chimpanzees make and use tools to solve a great range of problems. Primary among these at Gombe is the "fishing" of termites from underground nests with the aid of a stem of grass or a twig. In the *National Geographic* article "My Life Among Wild Chimpanzees," Jane describes this most important of her discoveries: "Termites form a major part of the chimpanzee diet for a two-month period. The termite season starts at the beginning of the rains, when the fertile insects grow wings and are ready to leave the nest. At this time, the passages are extended to the surface of the termite heap and then sealed lightly over while the insects await good flying weather. The chimpanzee is not alone in his taste for termites—the baboon in particular has a fondness for the juicy insects, but he must wait until they fly and then take his turn, together with the birds, at grabbing the termites as they leave the nest.

"The chimpanzee forestalls them all. He comes along, peers at the surface of the termite heap and, where he spies one of the sealed-off entrances, scrapes away the thin layer of soil. Then he picks a straw or dried stem of grass and pokes this carefully down the hole. The termites, like miniature bulldogs, bite the straw and hang on grimly as it is gently withdrawn.

"As the straw becomes bent at the end, the chimpanzee breaks off the bent pieces until the tool is too short for further use. Then it is discarded and a new one picked. Sometimes a leafy twig is selected, and before this can be used the chimpanzee has to strip off the leaves.

"In doing so—in modifying a natural object to make it suitable for a specific purpose—the chimpanzee has reached the first crude beginnings of tool making."

Chimpanzees in different parts of Africa show different tool-using behaviors. It is clear that infants learn these traditions by observing the behavior of adults, then practicing. One definition of human culture is behavior that is passed from one generation to the next through observation and imitation—thus we can say that chimpanzees have primitive cultures. At Gombe they use sticks for a variety of purposes: to fish for the vicious biting siafu (army ants), to enlarge holes in trees to search for honey, to investigate objects they can't touch or that provoke fear (the end of the probe is then sniffed).

They also use leaves as tools. When chimpanzees find water in a tree hollow that they can't reach with their lips, they chew a few leaves, crumple them to make a sponge they can dip into the hollow, and then suck out the liquid. They also use leaves as napkins to dab at blood or wipe away dirt from the body. And, they use sticks and rocks as weapons—clubs or missiles—during aggressive interactions with other chimpanzees, baboons, or humans. In West Africa, chimpanzees use rocks as hammers and anvils to smash open hard-shelled nuts, and one group was observed modifying branches to create rudimentary spears, which they thrusted into hollows in trees where lesser bushbabies sleep.

BELOW: Young Glitta (left) watches intently as her older sister, Gaia, fishes for termites.

## Family Groups

The individuals within a chimpanzee community, particularly family members, develop close, supportive, and affectionate bonds that can persist throughout their lives—for forty or fifty years. Chimpanzees maintain and strengthen relationships through long bouts of social grooming. They also cooperate during hunting and defending territories.

Wild chimpanzees become sexually mature when they are between ten and thirteen years old. Females are likely to mate with all or most of the males in their community, although never with their own sons. The father plays no role in childrearing, but the aforementioned patrols serve as a means of protecting resources for the females and young. Adult males are very protective of infants in their group. If a mother dies, her infant will almost always be adopted, usually by an older sibling—and brothers make excellent caregivers. Occasionally an unrelated adult will care for an orphan.

The chimpanzees' gestation period is approximately eight lunar months. A female seldom has her first baby before she is twelve years old. She then typically gives birth every five years or so. High infant mortality rates mean that she's unlikely to raise more than three offspring to full maturity during her lifetime.

A newborn chimpanzee infant is practically helpless but, unlike a human baby, has great strength in his hands and feet so he can cling to his mother, holding onto her hair as she moves about. Chimpanzee infants usually take their first steps at roughly four months, but they are very unsteady. They are totally dependent on their mothers for transport and milk for the first three years. Young chimpanzees are usually weaned at five years, but stay with their mothers until they are at least seven.

Infants are very active and adults are usually extremely tolerant of small infants playing around them. "I once watched little Fifi tormenting an adolescent male, Figan [her brother]," Jane wrote in a *National Geographic* article. "He was resting peacefully when Fifi hurled herself onto him, pulling his hair, pushing her fingers into his face, biting his ears. She swung above him, kicking out, while he indulgently pushed her to and fro with one hand. Finally, exhausted for the moment, she flung herself down beside him."

For the male, puberty begins at eight years of age and, as with humans, adolescence can be stressful. A young male gradually gains status with the females of his group as he gets older, but is likely to continue to show respect for his mother. But even as he gains dominance over the females, the adolescent male must learn to show increasing deference toward the older males. Behavior they tolerated when he was a mere juvenile may now be seen as a threat and he is likely to be firmly put in his place. In this time of uncertainty, an adolescent male often loses confidence and may spend greater amounts of time alone or with his mother. While male chimpanzees will remain in their natal groups throughout their lifetimes, adolescent females often join other communities.

ABOVE: Two youngsters frolic in the trees.

OPPOSITE, TOP: Gremlin sits with twins Glitta and Golden. They are the only Gombe twins known to have survived into adulthood.

OPPOSITE, BOTTOM: Minutes after her infant is born, this mother chimpanzee in the Republic of the Congo tenderly caresses the baby's tiny foot.

FOLLOWING PAGES: Gremlin rests with twin daughters Golden and Glitta, while her daughter Gaia grooms her. Behind Gremlin sits her son Galahad.

Young females shape their mothering skills by watching others and by grooming, carrying, and playing with younger siblings—or the infants of other females, if this is allowed. It became obvious early on in Jane's study of the Gombe chimps that the mothering skills of female chimpanzees can vary greatly. Flo and another female, Passion, provided an interesting contrast. Both mothers were high-ranking females, but had very different personalities. Flo was sociable and had relaxed relations with adult males. She was an attentive, affectionate, and supportive mother. Passion was asocial and her relations with community males were tense. She was a much less caring mother than Flo and much less affectionate. She seldom played with her daughter, Pom. Probably as a result of her favorable early experience, Fifi grew up to be well integrated in her community, and had successful offspring of her own. Pom, by contrast, was tense and nervous in her interactions with others and lost her first infant. After Passion died, Pom emigrated from her community and disappeared.

Within each chimpanzee community, there is an adult-male hierarchy in which each knows his place. And almost always, one male is clearly the number one, or "alpha." Usually, females are dominated by adult males. The dominance hierarchy of the females is less clear-cut, but some are obviously dominant and some clearly low-ranking. Although males may fight fiercely to attain the alpha position, this elevated status does not guarantee exclusive mating rights or the best food, but it does mean that all the others show deference.

YOUNG FEMALES SHAPE THEIR MOTHERING SKILLS BY WATCHING OTHERS AND BY GROOMING, CARRYING, AND PLAYING WITH YOUNGER SIBLINGS.

# CHIMPANZEE FACTS

DURING FIFTY YEARS STUDYING FREE-RANGING CHIMPANZEES AT GOMBE,
JANE GOODALL, HER STUDENTS, AND FIELD STAFF HAVE OBSERVED ALL ASPECTS
OF BEHAVIOR AND THE LIFE CYCLE FROM BIRTH TO DEATH.

## GENERAL

• Chimpanzees show intellectual abilities once thought unique to humans.
• Like us, they experience emotions such as joy, sadness, fear, and despair.
• They can be infected with all human contagious diseases (except cholera)—and we can catch theirs.
• The structure of their DNA differs from that of humans by just over 1 percent.
• Humans and chimps can exchange blood (if the blood groups are matched).
• Unlike humans, chimps do not swim or cry tears.

## FEEDING

• Chimpanzees, like humans, are omnivorous, eating meat, eggs, and insects, as well as fruits, leaves, shoots, blossoms, stems, and bark.
• They are successful hunters, sometimes showing sophisticated cooperation.
• They often eat mouthfuls of leaves with each bite of meat, as we eat vegetables.
• They rarely show cannibalistic behavior.

## TOOLS

• Chimpanzees use many objects as tools, and also modify them if necessary:

**grass stems** or twigs stripped of leaves to extract termites from their nests

**sticks** to feed on army ants and investigate things they cannot or do not want to touch

**leaves** to wipe dirt off themselves or, crumpled as sponges, to sop up water they cannot reach with their lips

**rocks and branches** as weapons

## SHARING

• Most sharing among chimpanzees occurs between mothers and infants. But adults will share meat after a successful kill, in response to begging.
• One adolescent female, on three occasions, climbed down from a tall tree with food in her mouth and hand to feed her old, sick mother.

## COMMUNICATION

• Chimpanzees have a rich repertoire of calls, each in their own distinct voice.
• Many of their postures and gestures are uncannily like some of our own: When greeting, chimpanzees embrace, hold hands, kiss, and pat one another. This helps calm excited or nervous individuals.
• Aggression among chimpanzees includes threats (waving arms and swaggering) and attacks (punching, kicking, stamping, biting). Females sometimes scratch and pull each other's hair.
• Disputes between chimpanzees are often solved by threats.
• Attacked victims often adopt a submissive, appeasing posture, which usually triggers reassurance—patting, embracing, or kissing—from the aggressor, restoring social harmony.
• Long sessions of social grooming serve to reinforce friendships and provide relaxation.

## LIFE STAGES

• A chimpanzee's first tooth appears at about three months. Permanent teeth start coming in during the fifth year.
• Solid food is not an important part of the chimpanzee diet until the age of three.
• Chimpanzees experience a long period of infant dependence on the mother. An infant suckles, rides the mother's back, and sleeps in the night nest during the

first five years, or until the next infant is born.
• Male puberty is reached at nine years old, and social maturity around age thirteen. Adolescence can be a frustrating time for males, who become more aggressive toward females but must be cautious with adult males, who fascinate yet frighten them.
• The female develops her first small swellings of the sex skin when she is about eight and is mated by infant and juvenile males. But she is not interesting to the mature males until she is about ten years old. In Gombe females don't conceive until eleven or twelve.
• Wild chimpanzees begin to look old when they are about forty-five years old. Captive chimps have lived as long as sixty-five years.

## SOCIETY
• Chimps live in a complex society. All fifty or so members of the community know each other as individuals.
• Theirs is a male-dominated society in which adult males compete for top rank (alpha male) and may reign for as long as ten years.

• Chimpanzees are aggressively territorial.
• Males patrol boundaries regularly and will attack "stranger" males and females, who often die of their wounds.

## REPRODUCTION
• Females generally give birth to one infant at a time. Only four sets of twins are known to have been born in Gombe since Jane's research began; three sets to the "G" family.
• There is an interval of about five years between live births. If an infant dies, the mother can become pregnant again within a couple of months.
• Pregnancy is eight lunar months.
• A sexually receptive female might be mated by males one after the other. Some males show possessive behavior and try to inhibit other males from mating, or a male will persuade or force a female to follow him on a "consortship," when he keeps her to himself.
• Infant mortality is high, and a female is unlikely to raise more than three offspring to full social maturity during her lifetime.
• Excessive inbreeding is avoided because of two factors: the first being

that females often transfer to another community before breeding and the second being a sexual inhibition—like an incest taboo—that typically prevents mating between mother and son and, to some extent, between brother and sister and even father and daughter.
• Some females have more sex appeal than others. Often an old and experienced female is more popular than a young and nervous female.

## THE FAMILY
• The mother is responsible for raising her infants. There are good chimpanzee mothers and bad ones. Most, however, are extraordinarily patient, tolerant, affectionate, and playful.
• All males show protective behavior to infants in their own communities.
• A sister or brother will play with, groom, and help protect a new sibling.
• Bonds between family members are close, affectionate, supportive, and may last through life.
• After the death of a mother, her infant—even though nutritionally independent—may be unable to recover from the trauma and may die.
• Older siblings typically adopt their young brothers or sisters if the mother dies. Males can be very competent caregivers; however, if the orphan is under three years and dependent on mother's milk, it will die anyway.

ABOVE, TOP: Frodo is bristled and tense during a conflict. A few months later he became the alpha male of the Kasekela community.

ABOVE, BOTTOM: Titan pant-hoots.

OPPOSITE: Jane pant-hoots with orphan Uruhara, who was rescued, emaciated and nearly hairless, by the Jane Goodall Institute and brought to the Sweetwaters Sanctuary.

## Communication

Relationships among adult chimpanzees are complex. Prior to Jane's studies, it was assumed that chimpanzees lived in harems, with a single alpha male and several females. In fact, they live in social groups within which each individual knows each other; but while some spend much time together, others meet but seldom. The small, scattered groups of a community maintain contact with each other through their calls.

Chimpanzees use a variety of sounds, each of which expresses a specific emotion and is understood by those who hear it. Calls range from the rather low-pitched *hoo* of the worried individual, the soft pant-grunts of greeting, to the loud, excited calls and screams that occur when delicious food is found or when two groups meet. One call, given in defiance of a possible predator or during aggressive interactions, can be described as a loud *wraaaah*. This is a single, drawn-out syllable, several times repeated, and can be a savage and even spine-chilling sound. Another characteristic call is the pant-hoot—a series of hoots, the breath drawn in audibly after each hoot, and often ending with three or four roars. There are several variations on this theme, but its most important function is to identify the caller, for the pant-hoot of each chimpanzee is distinctive. Hearing a pant-hoot, an individual can decide whether to reply, return the vocalization, or hasten away.

Chimpanzees also communicate by touch and gesture. A mother will touch her child when she is about to move away, and may tap on a tree trunk when she wants the youngster to come down. A chimpanzee who wants food will hold out a hand—palm up—in a humanlike gesture of begging. Friends greeting each other after a separation may embrace, kiss, or pat one another on the back. Less friendly individuals may swagger and wave their fists in the air. Angry or alarmed individuals will always erect the body hair, which is the effect you would get if a slightly built human male, with a bare torso, could cram two years of bodybuilding into a second.

Males sometimes perform spectacularly at the start of a heavy rainstorm, charging rhythmically across the ground, slapping and stamping, swaying vegetation, throwing a branch. In her book *In the Shadow of Man*, Jane describes the first time she saw a group of males performing their "rain dance": "At about noon the first heavy drops of rain began to fall. The chimpanzees climbed out of the tree and one after the other plodded up the steep grassy slope toward the open ridge at the top. There were seven adult males in the group, including Goliath and David Greybeard, several females, and a few youngsters. As they reached the ridge, the chimpanzees paused. At that moment the storm broke. The rain was torrential, and the sudden clap of thunder, right overhead, made me jump. As if this were a signal, one of the big males stood upright and as he swayed and swaggered rhythmically from foot to foot, I could just hear the rising crescendo of his pant-hoots above the beating of the rain. Then he charged off, flat-out down the slope toward the trees he had just

left. He ran some thirty yards, and then, swinging round the trunk of a small tree to break his headlong rush, leaped into the low branches and sat motionless.

"Almost at once two other males charged after him. One broke off a low branch from a tree as he ran and brandished it in the air before hurling it ahead of him. The other, as he reached the end of his run, stood upright and rhythmically swayed the branches of a tree back and forth before seizing a huge branch and dragging it farther down the slope. A fourth male, as he, too, charged, leaped into a tree and almost, without breaking his speed, tore off a large branch, leaped with it to the ground, and continued down the slope. As the last two males called and charged down, the one who had started the whole performance climbed from his tree and began plodding up the slope again. The others, who had also climbed into trees near the bottom of the slope, followed suit. When they reached the ridge, they started charging down all over again, one after the other, with equal vigor.

"The females and youngsters had climbed into trees near the top of the rise as soon as the displays had begun, and there they remained watching throughout the whole performance. . . . I could only watch and marvel at the magnificence of those splendid creatures. With a display of strength and vigor such as this, primitive man himself might have challenged the elements."

Threatening gestures and calls are more frequent in chimpanzees than are actual physical fights. When fights do break out, the most common causes are competition for status, defense of family members, and frustration that leads a thwarted individual to vent his aggression on a smaller or weaker bystander. Fights also may break out over food or access to a female.

## Territorial Aggression

A chimpanzee community has a home range within which its members roam in nomadic fashion. At Gombe, the home range of the main study community has fluctuated between five and nineteen square kilometers. The adult males, usually in groups of three or more, quite regularly patrol the boundaries, keeping close together, silent, and alert. If the patrol meets up with a similarly sized group from another community, both sides, after exchanging threats, are likely to withdraw discreetly back into home ground. But if a patrol meets a single individual or a mother and child, then the patrolling males usually chase and, if they can, attack.

In the early 1970s, the main study community at Gombe began to divide. Seven males and three females with offspring established themselves in the southern (Kahama) part of the original home range. During the next two years, these individuals returned to the north less and less frequently. Eventually they completely separated from the main group.

For a time the situation seemed fairly peaceful. If groups of the northern

(Kasekela) and southern communities met near their common boundary, the males would display, calling loudly, drumming on the trees, dragging branches as they charged back and forth. These displays served to persuade members of both groups to turn back into their respective home ranges.

Then, in early 1974, violence broke out between the two groups when five chimpanzees from the Kasekela community caught a single male of the Kahama group and brutally attacked him. He subsequently died of his wounds. The Kasekela males repeated these attacks again and again. Even Kahama females did not escape this violence. By the end of 1977, five males and one female had been killed and the remaining adults had disappeared. Only the adolescent females had escaped the violence. The victorious males tried to recruit them, but failed. The researchers at Gombe had observed a phenomenon rarely recorded in field studies: the gradual extermination of one group of primates by another, stronger group. Over the years, chimpanzees in all three communities in Gombe continued to vie for territory, with sometimes fatal consequences.

While such brutality is disturbing, Jane is quick to point out that chimpanzees are also capable of altruism. For example, two infants, Mel and Darbee, each about three and a half years old, were orphaned by a pneumonia epidemic. Both orphans were at first adopted by unrelated adolescent males, Spindle and Beethoven, who had lost their own mothers. Spindle would even share his night nest and allow Mel to ride, clinging to his belly, if it was rainy and cold. Later, both orphans were taken on by a childless female, Gigi.

Through the years, Gombe researchers have continued to look at chimpanzee feeding behavior, ecology, infant development, and aggression. They have also documented details of chimpanzee "consortships"—when males manage to lead females away to the periphery of the home range. Once there the male can relax without fear of competition for mating. Jane suggests that chimpanzees thus show a latent capacity to develop more permanent bonds similar to monogamy or, at any rate, serial monogamy.

BELOW, LEFT: Two forest monitors from the Greater Gombe Ecosystem and the Masito-Ugalla Ecosystem are training to use Google Android mobile phones to collect GPS observations of the status and trends of their village forests, wildlife, and threats.

BELOW, RIGHT: Dr. Lilian Pintea (left), the JGI director of conservation science, with a representative of the Mwamgongo village near Gombe's northern boundary and Amani Kingu of JGI-TACARE, assesses the potential usefulness of one-meter satellite images for participatory mapping of village landscapes in Africa.

OPPOSITE, TOP: Fifty years after Jane's landmark observations of chimpanzees fishing for termites, researchers are still studying insect-eating at Gombe. Here, graduate student Robert O'Malley of the University of Southern California holds a vial of siafu ants, another chimpanzee food, which were collected as part of an insect survey.

OPPOSITE, BOTTOM: During her early years at Gombe, Jane drew sketches in her notebook to illustrate chimpanzee termite-fishing; today, O'Malley uses a video camera to document differences in style and efficiency between individual chimps.

# Research Today

After a half century of research, a synthesis of the life of the Gombe chimpanzees has emerged. It validates and greatly expands Jane's early observations. For while Jane started her research with little more than a pencil, notebook, and binoculars, sophisticated new technology enables today's researchers to collect more detailed data. These new tools include the use of laptops in the field (for quicker and more comprehensive analysis of data collected), global positioning system handsets, and geographic information system software, all of which enable accurate mapping of chimp ranges and natural resources. Satellite imagery allows measurement of habitat types and their changes over time.

In addition, noninvasive sampling of urine and dung can measure sex hormones, stress hormones, SIVcpz (a virus almost certainly ancestral to HIV), and signs of other infections. Fecal samples can provide enough DNA to confirm paternity and other genetic relationships, allowing us to explore new questions, such as whether fathers and their biological offspring enjoy special relationships (avoiding mating or behaving as allies, for example).

New research methods allow something else as well. We can now look outward to survey the whole of Gombe National Park, documenting and better understanding how human communities and chimpanzees compete for space. The clear task at hand: Trying to achieve ecological balance.

Caretakers at chimpanzee sanctuaries often become surrogate family members to orphans who were stolen from their mothers—providing physical comfort, care, and socialization.

# A NEW VISION

THERE WERE ONCE BETWEEN ONE AND TWO MILLION CHIMPANZEES, SOME LIVING

IN THE FORESTS OF TANZANIA, BURUNDI, AND UGANDA, WITH THE MOST SIGNIFICANT

NUMBERS IN THE VAST RAIN FORESTS OF THE CONGO BASIN AND INTO WEST AFRICA.

BUT LOGGING COMPANIES INVADED THE CHIMPANZEES' HABITAT, OPENING UP AND

BUILDING ROADS THROUGH PRISTINE FORESTS. SLOWLY BUT STEADILY, THE

CONTINENT'S EVER-GROWING HUMAN POPULATION TOOK OVER THE VIRGIN LAND.

# THEY CHOPPED TREES FOR FIREWOOD, CLEARED UNDERGROWTH FOR SUBSISTENCE FARMLAND, HUNTED FOREST ANIMALS TO SELL THEIR MEAT AT MARKETS, AND TOOK LIVING INFANT APES—GORILLAS AND BONOBOS, AS WELL AS CHIMPANZEES—FROM THEIR DEAD MOTHERS' ARMS TO SELL FOR ENTERTAINMENT, MEDICAL RESEARCH, OR AS PETS. TODAY, THERE ARE PERHAPS AS FEW AS THREE HUNDRED THOUSAND CHIMPANZEES LEFT IN THE WILD. MANY OF THESE ARE IN SMALL, FRAGMENTED PATCHES OF FOREST, WITH LITTLE HOPE OF LONG-TERM SURVIVAL.

ABOVE: Farm plots on the steep, deforested hillsides north of Gombe National Park

BELOW: Two infant chimpanzees in burlap containers that were confiscated from smugglers in Uganda

OPPOSITE: A logging truck hauls cut trees and bags of bushmeat from deep within an African forest. Logging roads have opened the forest interiors to poachers and bush-meat hunters.

PREVIOUS PAGES: A caretaker holds an orphan chimpanzee at the Ngamba Island Sanctuary.

When Jane realized the extent to which chimpanzees across Africa were endangered, she made a decision that would change her life forever. She would leave the solitude and beauty of the Gombe forest and the chimpanzees that she loved almost like family. Using the fame and recognition she had gained from her *National Geographic* articles and documentaries, Jane would address government officials and conferences to raise awareness of the plight of our closest living relatives. And she would seek ways to raise money to help in the fight to save them. Jane's home in Gombe became not a base but a brief getaway between tours, a place to replenish her energy and gain inspiration to continue her campaign to help the chimpanzees and protect their habitat. Though she still monitors the research, Jane had to relinquish the day-to-day responsibilities to others.

By 1994, Gombe was under siege from the growing village populations that surround it on three sides. The lush green forest stood out like a tiny island amid the

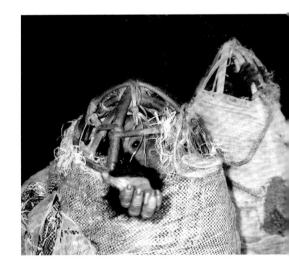

barren hills, the trees long ago cut by villagers in search of virgin soil for cultivation and building materials. Without the protection of trees, the precious topsoil washed down the steep hills into the lake, stripping the ground of fertility, polluting the once-clear waters, and endangering families with the terrible threat of mudslides.

Clearly, to be effective, any effort to save Gombe would have to address poverty and other deeply rooted human problems. As Jane puts it, "How can we even try to save the chimpanzees and forests if the people are so obviously struggling to survive?" Under the auspices of JGI, she initiated the TACARE (pronounced "take care") program, and a powerful force for conservation was born.

TACARE's community-centered conservation activities seek to preserve and restore the environment while helping villagers meet basic needs, such as education, health care, clean water, and arable land. Jane insisted on a participatory model, and that has defined the program throughout its development. The villagers were approached not by white people from outside, but by a team of carefully chosen Tanzanians familiar with the area. The TACARE staff asked the communities to identify their most urgent needs. Right from the start, the local people bought into TACARE. "By linking conservation to poverty alleviation," Jane says, "we've seen tangible results—whether a new forest plot or a mother newly able to afford school uniforms and books for her children." The program has helped create hope for thousands of families around Gombe.

Since 2003, JGI has initiated similar projects in the largely pristine Masito-Ugalla Ecosystem south of Gombe National Park, in the Maiko-Tayna-Kahuzi-Biega Landscape in the eastern Democratic Republic of the Congo, at the Tchimpounga Reserve in the Republic of the Congo, and at JGI's new Mount Otzi Program in Uganda. Supporters of JGI's efforts include the U.S. Agency for International Development, foundations, as well as other donors with in-kind technological contributions.

In 2006, JGI built on twelve years of experience with the TACARE program to design a larger and more comprehensive conservation initiative focused on restoring forest around Gombe. The Greater Gombe Ecosystem program blends TACARE community projects with conservation and land-use planning facilitated by geographic information systems mapping and analysis. It's an exciting initiative: JGI scientists meet with villagers, and together they pore over satellite maps and discuss how the forest has dwindled over the decades, how wildlife and people use the land now, and what might be the best use of given areas—including conservation—in the future.

ABOVE: A woman waters plants in a TACARE tree nursery near Kigoma, Tanzania.

OPPOSITE, LEFT: A villager waters seedlings in his tree nursery in the village of Zashe, Tanzania. As part of a TACARE project, he cares for the plants that have been given to him and is able to sell them when they are ready for planting.

OPPOSITE, RIGHT: Schoolchildren near the Ngamba Island Chimpanzee Sanctuary in Uganda

OPPOSITE, TOP: A woman uses a fuel-saving stove, as she was taught to do by TACARE staff.

OPPOSITE, BOTTOM: Now that TACARE has installed a sanitary water system in the village of Kasuku, women such as Amena Hassan no longer spend up to eight hours a day fetching water.

RIGHT: Dario Merlo, director of JGI's projects in the North Kivu province in the Democratic Republic of the Congo, shows pictures of chimpanzees to children in the village of Kasugho.

BELOW: Coffee beans are drying in the sun. Small-scale farmers who live on Gombe's border are encouraged to grow a coffee-bean variety that prefers shade, thus preserving the forest canopy and chimpanzee habitat. Through assistance from JGI TACARE, the beans were chosen by the Green Mountain Coffee Company for their Gombe Reserve Coffee.

# TACARE

**THE KIGOMA REGION OF WESTERN TANZANIA HAS THE
SECOND-HIGHEST YEARLY DEFORESTATION RATE IN THE COUNTRY.**

The shocking deforestation seen along the eastern shore of Lake Tanganyika from Kigoma to the Burundi border (and beyond) began in the 1970s, but was first noticed by Jane in the early 1990s. Suddenly, it seemed, the trees were gone. The presence of thousands of refugees from Burundi and the Democratic Republic of the Congo, along with the rapidly growing local population, were putting an insupportable demand on the natural resources, as forests were cleared for farming, building, and domestic uses.

JGI established the TACARE project in October 1994 to provide alternate and improved means of survival for people who live on the margins of the forest. Through education, micro-credit opportunities, health care, and support, TACARE helps women of the Kigoma region provide a better future for their families—and, in turn, a more promising future for the environment.

## THE GOALS OF TACARE

- Arrest the rapid degradation of land in the Kigoma region
- Improve the standard of living of the villagers by providing training and resources for growing fruit trees and other high-value crops
- Promote reforestation
- Curb soil erosion
- Provide conservation education to the local population
- Improve skills, education, and self-esteem of women
- Provide primary health care, AIDS education, and family planning services, in cooperation with regional medical offices

# NEW TOOLS FOR CHIMP RESEARCH AND CONSERVATION

## CHIMPANZEE POPULATIONS SUCH AS GOMBE'S MUST COEXIST WITH GROWING HUMAN POPULATIONS IN A SHARED LANDSCAPE.

Careful land-use planning can help bridge the two sets of needs. It can promote long-term survival of chimpanzees and better lives for impoverished human communities. But how can we ensure land-use plans are based on relevant, objective, and up-to-date information such as the actual chimpanzee distribution, habitat use, and ranging patterns?

Geographic information systems (GIS) technology is an invaluable asset, allowing Jane's teams to map chimpanzee presence and predict potential distribution and abundance, survey natural resources, and track human threats, such as settlements, farms, and other land uses.

Coupled with high-resolution satellite images, GIS data and maps can give insight into habitat loss and human activities over time. "Satellite imagery, GIS, and GPS have revolutionized the way we collect, analyze, and present information on forests, chimpanzees, and land use at the landscape scale. The latest satellite technology, such as QuickBird, allows us to map every tree, house, and footpath," says Dr. Lilian Pintea, the Jane Goodall Institute's director of conservation science.

The images themselves can be brought right into villages to facilitate participatory mapping and recording of local knowledge. Local people have been able to map their houses, streams, footpaths, and sacred areas. One woman was able to identify not only her field but also the tree under which she places her baby in the shade while farming. At the same time, the satellite imagery and maps are providing a unifying geographic framework where scientific and traditional knowledge can be combined. "Satellite images offer a common language, a way to communicate and share different perspectives on landscapes with people on the ground," Lilian says.

The land-use plans that Jane's teams have developed with villages in Tanzania include Village Forest Reserves that are protected buffer zones where certain extractive activities are allowed, agricultural zones, grazing zones, and human settlement zones. One long-term outcome around Gombe will be a forest corridor linking the forest reserves; it will provide a buffer and allow chimpanzees to move between Gombe and other forest patches outside the park.

This work is at the leading edge of conservation science and practice. Goodall and Pintea have made presentations about JGI's remote-sensing and GIS projects to interested audiences around the world, including preeminent mapping/GIS forums and technology leaders such as ESRI, DigitalGlobe, and Google Earth Outreach—all of which have been major contributors to JGI's efforts.

Lake Tanganyika (Elevation 773 m)

Gombe National Park Boundary

Milundi Mountain 1,520 m

LEFT· A satellite image shows the deforestation outside of Gombe's borders.

OPPOSITE: TACARE program manager Mary Mavanza talks with village women about TACARE programs to improve health and promote sustainable development.

OPPOSITE: A woman in the village of Kasuku, Tanzania, at the inauguration of a TACARE tree-planting campaign

RIGHT: Jane Goodall standing beside George Strunden, the JGI vice president of Africa programs, who helped develop the idea for TACARE with Jane.

BELOW: Children in the village of Kasugho in the North Kivu province of the Democratic Republic of the Congo

FOLLOWING PAGES: A village in the Greater Gombe Ecosystem where JGI TACARE provides assistance

# Sanctuaries in Africa

During Jane's years at Gombe, she had heard of the many chimpanzees kept as pets in the homes of expats or sold at village markets. And in 1990, she saw firsthand a tiny, frightened orphan, stolen from his mother by the hunters who had killed her for bushmeat.

"For years I have talked about the pitiful plight of infant chimpanzees who are sold in native markets," Jane says. "Now I have seen this with my own eyes. He was about one and a half years old, tied, by a short piece of rope, to the top of a chicken mesh [cage]. The trees overhead cast a little shade, but it was swelteringly hot and he was apathetic, dehydrated, and sweating. When I bent over him, he reached a gentle hand to touch my face. It is not legal to sell chimpanzees in this way in Zaire [now the Democratic Republic of the Congo]—not without all the proper permits. Yet he was brazenly exhibited right outside the American Cultural Center (in Kinshasa, Zaire). We knew we could not leave him there. Nor could we buy him, thus encouraging the trade."

This young chimpanzee, named Little Jay, was to launch yet another arm of the Jane Goodall Institute—one that was tasked with working on behalf of captive chimps. Because it is illegal to sell chimpanzees in range countries, Jane's first step was to encourage the governments to enforce their laws. To do this, government officials must confiscate the chimpanzees, showing poachers and traders there is no money to be made in this illegal venture. But once the authorities have the chimpanzees in their custody, the chimpanzees need a home. Without the appropriate rehabilitation and socialization, a captured orphan chimpanzee can't be returned to the wild because it lacks the skills needed to survive—skills it would have learned from its mother. And because chimpanzees are also very social within their communities, but territorial when dealing with outsiders, a lone chimpanzee placed in an unfamiliar forest would surely be killed.

The solution, therefore, was to provide a safe haven, a sanctuary where chimpanzees could be rehabilitated and live in peace—in the company of other orphaned chimpanzees. With the cooperation of the Burundi government and the assistance of the U.S. ambassador to Burundi, Dan Phillips, and his wife, Lucie, JGI set up a halfway house in Burundi's capital, Bujumbura, for chimpanzees smuggled across the border (mostly from neighboring Zaire). Soon, twenty chimpanzees were crowded into the facility. The hope was to create an appropriate sanctuary once funds were raised. Unfortunately, the political instability of this troubled country repeatedly put sanctuary plans on hold, and in 1995, escalating violence forced the institute to evacuate the chimpanzees. All twenty were airlifted to Sweetwaters Chimpanzee Sanctuary, two hundred acres of riverine woodland in central Kenya, near the foot of Mount Kenya, where, at last, they could climb trees, build nests, and move freely.

In 1998 JGI opened the Ngamba Island Chimpanzee Sanctuary, which is managed now by the Chimpanzee Sanctuary & Wildlife Conservation Trust on Ngamba Island, a hundred-acre rainforest paradise in Lake Victoria near Entebbe, Uganda. The sanctuary is a model of its kind, with a thriving educational program that includes forest walks during which visitors interact with infant chimpanzees on short, guided treks through the forest.

BELOW: While visiting a market in the Congo, Jane saw a tiny infant chimpanzee tied to a cage in the hot sun. He looked nearly dead. Jane knew she needed to find a way to rescue him. With the intervention of the American ambassador and the minister of the environment, the chimp, whom she named Little Jay, was rescued. He became the first of hundreds of rescued chimps who now live in sanctuaries in Africa.

OPPOSITE: A caretaker at Tchimpounga escorts a group of juveniles to a forest patch where they will spend the day.

Another sanctuary, the Tchimpounga Chimpanzee Rehabilitation Center, is situated in a forested area of about sixty-five acres near Pointe-Noire in the Republic of the Congo. Built initially with the help of Conoco Inc., it was opened in 1992. The number of orphans arriving at the sanctuary, always traumatized and often wounded, has grown steadily. By 2010, more than 140 chimpanzees had been housed and cared for there. They ranged in age from less than one year to full-grown. Undoubtedly the most famous resident of Tchimpounga was Gregoire, whom Jane first met in 1990 when he was an emaciated, almost hairless chimpanzee who'd been imprisoned in a lonely cage in a Brazzaville zoo since 1944. Gregoire was transferred to the sanctuary in 1997 during the civil war, while heavy fighting was occurring in the capital. He lived the rest of his life at Tchimpounga and was at least sixty-five years old when he died in December 2008.

Jane also lent her support to the creation of a sanctuary and ecotourist site in South Africa: the Jane Goodall Institute Chimpanzee Eden. Set on a gorgeous thousand-hectare reserve, the sanctuary is home to about thirty orphaned chimps. It was the setting of an Animal Planet television series that highlighted the efforts of sanctuary founder Eugene Cussons to rescue the chimps and rehabilitate them into social groups. Chimp Eden has greatly aided Jane's efforts to educate the world about the plight of wild chimpanzees.

# TRIBUTE TO GREGOIRE

## "DURING MY FIRST VISIT TO BRAZZAVILLE ZOO IN CONGO, I MET GREGOIRE," JANE SAYS.

"I can still recall my sense of disbelief and outrage as I gazed at this strange being, alone in his bleak cement-floored cage. His pale, almost hairless skin was stretched tightly over his emaciated body so that every bone could be seen. His eyes were dull as he reached out with a thin, bony hand for a proffered morsel of food. Was this really a chimpanzee? Apparently so. Above his cage was a sign that read 'Shimpanse—1944.' 1944! It was hard to believe. In that dim, unfriendly cage, Gregoire had endured for forty-six years!

"A group of Congolese children approached him quietly. One girl, about ten years old, had a banana in her hand. Leaning over the safety rail, she called out, 'Danse! Gregoire—Danse!' With bizarre, stereotyped movements, the old male stood upright and twirled around three times. Then, still standing, he drummed rapidly with his hands on a single piece of furniture in his room, a lopsided shelf attached to one wall. He ended the strange performance by standing on his hands, his feet gripping the bars between us. The girl held the banana toward him and, righting himself, he reached out to accept his payment.

"That meeting was just after Nelson Mandela had been released from his long imprisonment by the white South African government. I was with a Congolese official at the time, who knew nothing of chimpanzees. After staring at Gregoire for a while he turned to me, his face solemn. 'There, I think, is our Mandela,' he said. I was moved by those words, by the compassion that lay behind them.

"The gaunt image of Gregoire hung between me and sleep that night. How had he survived those long, weary years deprived of almost everything that a chimpanzee needs to make life meaningful? What stubbornness of spirit had kept him alive? It was as though he, like other starving, neglected chimpanzees in impoverished African zoos, had been waiting for help."

At Jane's insistence, JGI hired a Congolese worker to look after Gregoire. Soon, his barren cage was expanded to include a small outdoor "deck" and he began to grow hair and put on weight. In late 1996, Gregoire was introduced to a four-year-old male orphan named Bobby and an infant female, whom Jane named Cherie. Despite the fifty-year age difference, Bobby and Gregoire played together like children. And the old male treated young Cherie like a much-loved granddaughter.

In May 1997, civil war broke out in Congo. Fortunately, a team of wildlife experts rescued the great apes of the Brazzaville zoo and moved them to safety. Gregoire and his companions were moved to JGI's Tchimpounga Chimpanzee Rehabilitation Center, where Cherie slept each night in Gregoire's arms.

One of the animal world's most incredible stories of resilience and happy endings came to a quiet close on December 17, 2008: Gregoire, Africa's oldest-known chimpanzee and a national hero in the Republic of the Congo, died in his sleep. As news of Gregoire's death circulated, messages of condolence and sympathy made their way to Jane and JGI from around the world.

Gregoire was nowhere better known than in Congo. "Everyone knew Gregoire," says Lisa Pharoah, JGI's West and Central Africa program manager. "Children ... adults ... they all had stories about him. You could tell people you worked for JGI and maybe you would get a reaction. But tell people you work with Gregoire? They'd get so excited: "Oh, Gregoire!"

OPPOSITE: Jane and Gregoire groom each other.

FOLLOWING PAGES: The 140 chimpanzees at the Tchimpounga Sanctuary consume nearly 1,400 pounds of fresh fruit and vegetables every day.

# EVERY YEAR, POACHERS KILL THOUSANDS OF CHIMPANZEES
AND OTHER ENDANGERED ANIMALS, RANGING FROM ELEPHANTS AND GORILLAS TO BIRDS AND BATS—ANYTHING THAT CAN BE SMOKED AND SOLD AS FOOD.

## The Commercial Bushmeat Trade

ABOVE: A severely traumatized infant chimpanzee arrives at the Tchimpounga Sanctuary with burns on her face.

OPPOSITE, TOP: Tchimpounga Natural Reserve ecoguards Christ (left) and Jerome are seen here burning confiscated bushmeat, as per Congolese law. The burning is necessary to ensure that those who find it do not profit from the bushmeat themselves.

OPPOSITE, CENTER: A JGI-Uganda veterinary intervention team—led by veterinarians Dr. David Hyeroba, Dr. Peter Apell, and Dr. Tony Kidega—removes a jaw-trap from the wrist of a wild adolescent female in the Rwensama forest in Uganda. The procedure was successful. She regained full use of her hands and went back in the forest with the rest of her group.

OPPOSITE, BOTTOM: This chimpanzee's wrist was caught in a trap set to capture bushmeat.

FOLLOWING PAGES: Local schoolchildren visit the chimpanzees at the Ngamba Island Sanctuary in Uganda to learn about the importance of protecting chimpanzees and the environment.

In the last twenty to thirty years, logging and mining companies in the Congo Basin have developed road networks in formerly pristine forests, giving poachers access to new territory. One result: a dramatic escalation in hunting of wildlife for meat. Every year, poachers kill thousands of chimpanzees and other endangered animals, ranging from elephants and gorillas to birds and bats—anything that can be smoked and sold as food. Baby chimpanzees generally aren't killed, but captured to be sold as pets or to attract visitors to hotels and bars.

The illegal commercial trade isn't driven by the need to feed local people. Instead, much bushmeat ends up in city markets and expensive restaurants throughout Africa. In great ape range countries, in Europe, and even in the United States, wealthy, elite diners regard fare such as chimpanzee hands as a delicacy.

JGI's Tchimpounga Chimpanzee Rehabilitation Center is in the heart of the Congo Basin's commercial bushmeat corridor. It has been estimated that every infant chimpanzee rescued from the black market and brought to Tchimpounga represents ten to twelve other chimpanzees killed by the trade.

Tchimpounga's orphaned chimps serve as ambassadors. When local people visit and see these amazing, charismatic beings up close, they realize how similar chimps are to humans. Many say they will never eat chimpanzee again.

In collaboration with the sanctuary, JGI-Congo offers formal environmental education for area communities and targeted schools, encouraging both adults and children alike to appreciate their surrounding forests and wildlife. JGI-Congo education staff members also conduct bushmeat public-awareness campaigns and workshops with police and other authorities, and with the women and men who sell bushmeat in markets. They can see attitudes slowly changing, reminding them of the value of this challenging endeavor.

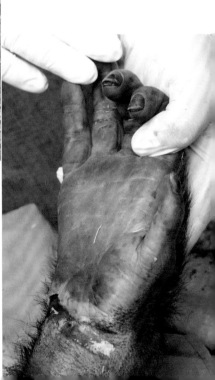

# TCHIMPOUNGA CHIMPANZEE REHABILITATION CENTER

## THE TCHIMPOUNGA CHIMPANZEE REHABILITATION CENTER WAS ESTABLISHED IN THE REPUBLIC OF THE CONGO IN DECEMBER 1992 WITH THE SUPPORT OF CONOCO INC.

Tchimpounga is run by a remarkable team made up of local caregivers, maintenance workers, and veterinary health and management staff led by Dr. Rebeca Artencia. Their dedication has been many chimps' only hope for survival amid Congo's civil wars and ongoing political strife.

An escalation of the bushmeat trade supplies an ever-increasing population of orphans; there are now more than 140 chimpanzees living at Tchimpounga, stretching it to maximum capacity. The sanctuary includes large enclosed forests and other outdoor areas that the chimpanzees access and explore daily, as well as dormitories which the chimpanzees use nightly (for sleeping) and which also allow for easy access and/or isolation if a chimp is in need of veterinary intervention.

Tchimpounga is situated on a coastal plain of savannah and galleried mosaic forest patches. It is located within the Tchimpounga Natural Reserve, thirty-one miles (fifty kilometers) north of the city of Pointe-Noire in the Kouilou region. Its mission includes chimpanzee and habitat protection, research, education and outreach, ecotourism, and involvement of local populations in sustainable development initiatives.

BELOW: A caretaker feeds infant orphans at the Tchimpounga Sanctuary.

# SNARES IN UGANDA

## THE LEADER OF HIS CHIMPANZEE GROUP, KIGERERE WAS MOVING THROUGH THE FOREST ONE DAY AND SUDDENLY TRIPPED A SNARE.

Hunters in Uganda set such snares for bushpigs, antelope, and other prey more abundantly than chimpanzees, but these illegal snares capture as many as 25 percent of chimps in Uganda. They're typically made of metal or wire and easily hidden in the forest underbrush. And they are viciously indiscriminate.

When an animal struggles to escape, it only tightens the snare. Some animals chew their own limbs off in a desperate attempt to free themselves.

Of course, Kigerere tried desperately to free himself. The chimps in his group brought him food and stayed by him as he struggled to survive. But Kigerere couldn't loosen the snare and he couldn't save himself.

The Jane Goodall Institute in Uganda works to prevent such miserable deaths for chimpanzees through a snare removal program that involves local community members, including former poachers. It makes sense to hire these men—they know best where to look for

snares and now have an alternative livelihood to poaching.

The project teams scour the forest floor and remove the illegal snares. They use handheld GPS systems to mark snare points, and then use that data to map out high-risk areas for snares. This facilitates future snare patrols.

JGI-Uganda gives the snares to local women to use in arts and crafts such as decorations on greeting cards—a source of income and a proactive way to spread the word about chimpanzees and poaching.

BELOW: Two forest guides in the Bundongo Forest Reserve in Uganda hold snares often used to trap and maim wild chimpanzees.

# ChimpanZoo

On the road, or flying from city to city, auditorium to auditorium, and meeting to meeting, Jane was often asked to visit chimpanzees in zoos, and in so doing she came upon an idea:

"I suddenly realized the exciting potential of these invaluable captive groups," Jane says. "If only, I thought, we could collect comparable data at a number of different zoo sites, using the same behavior categories as we do at Gombe, and similar recording techniques, a whole wealth of new information would soon become available. We would gradually learn more about the extent to which chimpanzee behavior is flexible. In how many ways do chimpanzees adapt to the different social and physical environments? It would give us a chance to tap into the vast store of knowledge and expertise so often stored away in the heads of individual keepers. And it would also lead to better environments for the chimpanzees."

In 1984, Jane founded ChimpanZoo, an international program dedicated to the well-being and understanding of chimpanzees in zoos and other captive situations. Volunteers, students, and sometimes keepers participate, collecting information that is shared through a database managed by JGI. Today the program has evolved to include zoos in Europe and other parts of the world. Approximately 130 chimpanzees are involved in ChimpanZoo, making it the largest captive-ape research program ever undertaken.

The program has created a network of zoos with the collective know-how to respond to almost any husbandry problem. Dr. Virginia Landau, the longtime director of ChimpanZoo, has provided a unique service to those seeking advice from around the world, answering questions about health, living conditions, introductions, birth control, and countless behavioral problems.

RIGHT: Kathy peers out from her indoor enclosure to the spacious outside yard at the Lincoln Park Zoo in Chicago, Illinois. Former Gombe researcher Dr. Elizabeth Lonsdorf leads the Lester E. Fisher Center for the Study and Conservation of Apes at the Lincoln Park Zoo, one of several institutions committed to the research and conservation of chimpanzees. The zoo partners with the Jane Goodall Institute on a number of research projects in Gombe National Park.

OPPOSITE: A docent explains the exhibit to visitors at the Lincoln Park Zoo.

ChimpanZoo participants gather data on the behavior of individual chimpanzees in all participating zoos. Data is collected systematically, using the same methodology across the board, and entered into a database to enable researchers to make use of all this information. Notes are kept on important events in the captive group—births, deaths, arrival or departure of individuals, changes in dominance, serious fights, changes in enclosure, changes in primary keepers, and so on. Standard questionnaires may be used to assess the personality type of each chimpanzee (as developed by Dr. James King). There is an annual week-long ChimpanZoo conference that serves as a forum for discussing and exchanging new information and ideas. It attracts the general public as well as professionals from the academic and zoological communities from around the world. Many of the papers presented have been published in scholarly journals, and the ChimpanZoo database is accessible to zoos, students, and instructors.

Finally, ChimpanZoo provides information about chimpanzee behavior to schoolchildren of all ages, especially through Jane Goodall's Roots & Shoots, and to their parents and teachers. Involvement of high school and first-year university students in data collection prepares them for higher-level participation. A ChimpanZoo course for undergraduates ran successfully at Colorado University for several years, and ChimpanZoo has developed excellent teaching materials for younger children.

## Chimpanzees as Pets and Entertainers

In separate incidents in the 2000s, two chimps—one a household "pet" and one a trained "entertainer"—savagely attacked humans known to them, giving the public an unforgettable lesson in the dark side of chimps. The message couldn't have been clearer: Chimps aren't meant to be pets or media props. Jane and her staff work hard to spread this message, using action-and-awareness campaigns, as well as direct appeals to the entertainment and advertising industries. Jane stresses one key point: Chimps may be adorable as babies, but they grow to be many times stronger than any human. They are, at their core, unpredictable and dangerous. Despite careful upbringing with humans, they remain wild. Once a chimp is no longer manageable, he or she has to be caged or sent away. Good zoos typically won't take a chimp that hasn't been properly socialized with other chimps. The poor creature may end up in a roadside "zoo," caged for the rest of its life, often alone. Or, he or she could wind up in a research lab. It's a terrible outcome; both for the original human owners, who often love the chimp dearly, and for the animal denied its birthright—life as a chimp.

## Medical Research Laboratories

Even as Jane became increasingly involved in the proper care of entertainment and pet chimpanzees, she was learning more about another deeply disturbing problem: conditions in medical-research laboratories. Though she could imagine the horrors these intelligent and aware creatures suffered in the small, bleak, and sterile cages of the research labs (five feet by five feet by seven feet per chimp at the time), Jane felt she could not speak out on their behalf until she had seen the conditions for herself. In 1987, Jane visited her first medical-research facility, and later wrote about the experience in *The New York Times*:

> *"Room after room was lined with small, bare cages, stacked one above the other, in which monkeys circled round and round, and chimpanzees sat huddled, far gone in depression and despair.*
>
> *"Young chimpanzees, three or four years old, were crammed, two together, into tiny cages measuring 22 inches by 22 inches and only 24 inches high. They could hardly turn around.*
>
> *"The chimps had each other for comfort, but they would not be together for long. Once they are infected, probably with hepatitis, they will be separated and placed in another cage. And there they will remain, living in conditions of severe sensory deprivation, for the next several years. During that time, they will become insane.*

*"A juvenile female rocked side to side, sealed off from the outside world behind the glass doors of her metal isolation chamber. She was in semidarkness. All she could hear was the incessant roar of air rushing through vents into her prison.*

*"I shall be haunted forever by her eyes, and the eyes of the other chimpanzees I saw that day.*

*"I have had the privilege of working among wild, free chimpanzees for more than twenty-six years. I have gained a deep understanding of chimpanzee nature. Chimpanzees have given me so much in my life. The least I can do is speak out for the hundreds of chimpanzees who, right now, sit hunched, miserable, and without hope, staring out with dead eyes from their metal prisons. They cannot speak for themselves."*

And so she did. Jane spoke with people from the National Institutes of Health and directors of laboratories that housed chimpanzees. She spoke with members of Congress. She soon found out that the lab she visited was one of the worst, that others had larger cages or group housing. But by studying how the chimpanzees lived in the wild and understanding their rich social life, Jane knew that even the best of the conditions in which they were kept were sorely inadequate. She and her associates fought to have the regulations changed, requiring all labs to provide more humane conditions. They created committees to write recommendations to the government that would address some of their concerns. But of course, Jane also felt that chimpanzees, so much like us, should not be used as living test tubes anyway.

Over the years, others took up the cause of lab chimps. Jane and these advocates have seen great advances. Most notably, several countries now ban the use of great apes in medical research, including the United Kingdom, Austria, Belgium, the Netherlands, Sweden, and New Zealand. Unfortunately, the United States hasn't followed suit. In fact, it is the last large-scale user of chimpanzees in research; but awareness and advocacy efforts have not been fruitless. In 2000, the government mandated life-long care in sanctuaries for lab chimps who are no longer considered useful. Then in 2007—thanks in part to the efforts of JGI—the government closed a loophole that would have allowed the chimps to be pulled out of retirement if deemed necessary. The new rules state that retired chimps never be put back into labs.

The proposed Great Ape Protection Act of 2009 would end all invasive research on chimpanzees and other great apes in the United States, and would also require the federal government to retire all great apes that it is holding for invasive research. It seems only a matter of time before the world ends the use of great apes in invasive research and sterile lab conditions once and for all.

Meanwhile, Jane continues the good fight. In 2008, she joined the Dr. Hadwen Trust in presenting a petition to the European Parliament, urging a ban on all animal testing. She also suggested a Nobel Prize to award research on development of alternatives to animal testing.

ABOVE: Jane Goodall visits the Laboratory for Experimental Medicine and Surgery in Primates (LEMSIP) in New York. The cages where chimpanzees lived were five feet wide by five feet deep and suspended above the floor, forcing the chimps to spend their lives stepping on bars rather than on a solid surface.

OPPOSITE, TOP: Jane comforts a young chimp who serves as a living prop for a beach photographer in the Canary Islands.

OPPOSITE, BOTTOM: This deformed and abused chimpanzee lies in a small cage at a biomedical laboratory in Kinshasa, Democratic Republic of the Congo. Jane is a leading advocate for the abolition of the use of all live animals in research, and, until that happens, the humane treatment of research animals.

# THE HOPE

SINCE 1986, DR. GOODALL HAS NOT STAYED IN ANY ONE PLACE LONGER THAN THREE WEEKS, AND SHE HAS BEEN ON THE ROAD SOME THREE HUNDRED DAYS A YEAR. WHEN NOT AT HER HOME IN ENGLAND, HER DAYS ARE SPENT ON AIRPLANES, TRAVELING TO SPEAKING ENGAGEMENTS, RECEPTIONS, AND PRESS CONFERENCES, AND LOBBYING FOR ENVIRONMENTAL CAUSES. IT WAS HEARTBREAKING FOR DR. GOODALL TO LEAVE HER PARADISE AT GOMBE, BUT SHE KNOWS THAT WHAT SHE IS DOING NOW IS MORE IMPORTANT.

**T**HROUGH JANE'S WORK WITH CHIMPANZEES, WE LEARNED THAT HUMANS ARE NOT AS DIFFERENT FROM THE REST OF THE ANIMAL WORLD AS WE ONCE BELIEVED, AND THAT WE ARE NOT THE ONLY BEINGS CAPABLE OF RATIONAL THOUGHT, OF EMOTIONS, AND OF MENTAL AND PHYSICAL SUFFERING. ONCE WE ACCEPT THAT ANIMALS DO INDEED SUFFER, WE MUST BEGIN TO SCRUTINIZE THE WAY IN WHICH WE TREAT ALL ANIMALS, HUMAN AND NONHUMAN ALIKE, AS WELL AS THE ENVIRONMENT IN WHICH THEY LIVE.

## Jane Goodall's Roots & Shoots

In celebration of thirty years of chimpanzee research, Jane invited a select group of students to her house in Dar es Salaam, Tanzania. After talking about chimpanzee behavior and other experiences in Africa, Jane spoke about the problems, such as poaching, dynamite fishing, and cruelty to animals.

"How do you feel when you see chickens carried upside down, or by their wings held behind their backs?" Jane asked the students gathered around her verandah. She was referring to the way in which chickens are typically taken to market in Africa.

The students responded with compassion. Although the chicken is going to die, to become someone's supper, they replied, while the animal still lives we should treat it with respect. That was the response Jane had hoped for—a recognition that every individual matters, human and nonhuman alike, and that every individual deserves our compassion.

The young people decided to go back to their schools and start clubs to help reduce animal cruelty and address environmental problems. And so, in 1991, Jane Goodall's Roots & Shoots was born. The name is symbolic. The first pale roots and shoots of a germinating seed look so tiny and fragile; it's hard to believe it can grow into a big tree. Yet there is so much life force in that seed that the roots can work their way through boulders to reach the water, and the shoot can work its way through cracks in a brick wall to reach the sunlight. Eventually the boulders and the wall—all the harm, environmental and social, that has resulted from our greed, cruelty, and lack of understanding—will be pushed aside. And hundreds and thousands of roots and shoots, representing the youth of the world, can solve many of the problems their elders have created for them.

Since the program's inception, Roots & Shoots groups have sprung up throughout Tanzania and in more than 120 countries around the world. Tens of thousands of young people from preschool through university are involved. Roots & Shoots members are young people determined to make a difference, prepared to roll up their sleeves and take action, to walk the talk. It is a youth-driven program: The

ABOVE: Students in the Kibale Environmental Education Program (K.E.E.P.) participate in a clean-up project. The program aims to educate primary school students about the Kibale National Park in Uganda, where they live.

OPPOSITE, TOP: Schoolchildren in the Republic of the Congo learn about chimpanzees and the importance of protecting them.

OPPOSITE, BOTTOM: Jane Goodall and Roots & Shoots members at the Singapore American School plant a nutmeg tree.

PREVIOUS PAGES: These children in Masindi, Uganda, have an active Roots & Shoots program at their school.

# ROOTS & SHOOTS MEMBERS ARE YOUNG PEOPLE DETERMINED TO MAKE A DIFFERENCE, PREPARED TO ROLL UP THEIR SLEEVES AND TAKE ACTION, TO WALK THE TALK.

ABOVE: A youngster at the Langley School in McLean, Virginia, is thrilled to meet her hero, Jane Goodall.

ABOVE, RIGHT: JGI's youth programs are dedicated to the belief that young people, when informed and empowered, can indeed change the world.

young people themselves discuss local problems and decide what they can do to try to solve them. They choose projects that show care and compassion for the human community, animals (including domestic animals), and the environment. Their projects vary depending on the nature of the problems and the age of the members, their culture, whether they are from inner-city or rural environments, and which country they come from.

Projects range from cleaning a stream in Oregon, planting trees in Tanzania, walking a neighbor's dog in Minnesota, visiting children in a hospital in South Africa, and enriching the lives of zoo chimps in Shanghai. But they all learn about the interconnectedness and interdependence of life on earth.

Jane has become increasingly excited by the energy of young people, the huge potential of the program, and the hope for the future. She sees that attitudes and generations-old practices are changing through the power of the world's youth.

"I think Roots & Shoots is probably the reason I came into the world," Jane says. "Yet I couldn't have done it without all those years with the chimpanzees and an understanding that led to a blurring of the line between 'man' and 'beast.' Children give me particular hope because they have more open minds. They aren't as set in their ways. Only if children grow up with respect for all living things will the planet have a chance for survival.

"We need to ensure a critical mass of young people who understand that money is not everything. We need money to live; we should never live for money. It is when these individuals move into adult positions of decision-making that the world will change."

# JANE'S REASONS FOR HOPE

"THE GREATEST DANGER TO OUR PLANET IS THAT WE LOSE HOPE," JANE SAYS. "WITHOUT HOPE, WE GIVE UP, STOP TRYING TO DO OUR BIT TO MAKE A DIFFERENCE." IN SPITE OF ALL THE PROBLEMS WE FACE TODAY, JANE'S PERSISTENT MESSAGE IS ABOUT THE IMPORTANCE OF HOPE. BELOW ARE JANE'S FOUR REASONS FOR HOPE.

### THE HUMAN BRAIN

Think of all we've accomplished as a species. We found cures for so many diseases; we flew men to the moon. And we have at last begun to face the problems that threaten our survival. Surely we can use our quite extraordinary problem-solving abilities, our brains, to find ways to live in harmony with nature, to develop more sustainable societies.

### THE RESILIENCE OF NATURE

My second reason for hope is the incredible resilience of nature. I have visited Nagasaki, the site of the second atomic bomb that ended World War II. Scientists had predicted that nothing could grow there for at least thirty years. But, amazingly, greenery grew quickly. One sapling actually managed to survive the bombing, and today it's a large tree, with great cracks and fissures, all black inside; but that tree still produces leaves. I've seen such renewals time and again, including animal species brought back from the brink of extinction.

### THE DETERMINATION OF YOUNG PEOPLE

My third reason for hope lies in the tremendous energy and commitment of young people around the world. I meet so many young people with shining eyes who want to tell Dr. Jane what they've been doing, how they are making a difference in their communities. Whether it's something simple like recycling or collecting trash, something that requires a lot of effort, like restoring a wetland or a prairie, or whether it's raising money for the local dog shelter or victims of an earthquake in a faraway country, they are getting things done. Young people—their spirit and determination—are my greatest source of hope. We should never underestimate the power of determined young people to bring about change. We only have to give them the knowledge and the tools.

### THE INDOMITABLE HUMAN SPIRIT

My final reason for hope lies in the indomitable nature of the human spirit. There are so many people who have dreamed seemingly unattainable dreams and, because they never gave up, achieved their goals against all odds, or blazed a path along which others could follow. As I travel around the world, I meet so many incredible and amazing human beings. They inspire me. They inspire those around them.

OPPOSITE: Jane celebrates the International Day of Peace with Roots & Shoots members at the United Nations Headquarters in New York City.

ABOVE: Villagers in the Democratic Republic of the Congo embody the indomitable human spirit—one of Jane's four reasons for hope.

OPPOSITE, TOP: Roots & Shoots members take action all over the world through service projects that are as varied as young people's imaginations. Here, Jane attends a Friendship Meeting with children and teachers from the Ryul Gok Middle School in Pyongyang, North Korea.

OPPOSITE, BOTTOM: U.S. Roots & Shoots Youth Leadership members David Shorna (center) and Alessandra Phelan-Roberts (right) help a Roots & Shoots friend plant a tree in Moshi, Tanzania.

PREVIOUS PAGES: When Jane became a United Nations Messenger of Peace, she encouraged Roots & Shoots members everywhere to make giant peace doves and fly them in support of the UN's International Day of Peace. Each year the doves fly in countries around the world. Here, a dove soars near the Teton Mountains in Jackson Hole, Wyoming.

# SPREADING THE MESSAGE

THROUGH THE YEARS, PRODUCTION PARTNERSHIPS WITH NATIONAL GEOGRAPHIC,
DISCOVERY, AND OTHER COMPANIES NOT ONLY MADE JANE GOODALL A HOUSEHOLD NAME,
THEY HELPED HER REACH EVER-LARGER AUDIENCES THROUGH HER RECOUNTINGS
OF THE WONDERS OF CHIMPANZEES, HER REASONS FOR HOPE, AND, MOST IMPORTANTLY,
HER CONVICTION THAT EACH INDIVIDUAL CAN MAKE A DIFFERENCE.

The big screen has been Jane's ally as well. In October 2002, a giant-screen film, *Jane Goodall's Wild Chimpanzees*, premiered at Discovery Place in Charlotte, North Carolina. Since then, the film has played in eighty-three venues, reaching an estimated 3.24 million people. The first and only giant-screen film on Jane, it blended her incredible life story with archival footage and photos, vibrant African music, and close-up depictions of the Gombe forest and its chimpanzees. (Directed by Dave Lickley and written by Stephen Low, it was a co-production of the Science Museum of Minnesota, Science North, and Discovery Place, in cooperation with the Jane Goodall Institute.)

*Jane's Journey* is the first major motion picture about Jane. Director Lorenz Knauer and his film crew accompanied Jane on her travels across several continents. They captured the rigors and joys of her life on the road—the nonstop interviews, her ability to bring audiences to tears with her stories of hope and heroism, and a kaleidoscope of experiences. These notable moments include an unforgettable trip to Greenland where Jane could see firsthand the melting of the great glacier, and a moving visit with Roots & Shoots youths in the Lugufu refugee camp in Tanzania.

*Jane's Journey* promises to widen the community of people Jane has inspired to leave a softer footprint on the earth.

# Fifty Years

It's been fifty years since Jane Goodall first set foot on a lakeshore in Tanzania to begin her study of chimpanzees. With patience and determination, she managed to open a window onto the daily lives and dramas of our closest living relatives and was rewarded with the public's passionate interest in her subject. When the dire plight of chimpanzees became apparent, she knew she had to try to save them. Today, Jane and her staff at the Jane Goodall Institute are determined that your grandchildren will be able to see chimpanzees thriving in the wild. They work with an array of global partners to preserve chimpanzees—and their precious forest habitat—in concert with efforts to reduce poverty and to help African communities meet basic needs such as education, health care, clean water, and arable land.

Jane and her staff also support youth action, through the global Roots & Shoots program, because ultimately young people are the ones who will have the task of restoring vitality to the damaged environment. Informed and empowered youth have the optimism, the idealism, and the energy.

As Jane's global profile continues to grow (she became a United Nations Messenger of Peace in 2002, was named a Dame of the British Empire in 2004, and in 2006 received the French Legion of Honor), global offices of the Jane Goodall Institute sprout up around the world—residing in more than twenty-five countries as of 2010. These offices, some staffed entirely by volunteers, all work together to support Jane and help realize her vision for a better world.

It is an amazing legacy for one who, in her youth, simply dreamed of living with the animals and writing books about them. Thank goodness Jane never gave up on that dream. The chimpanzees face many challenges, but they could not ask for a more determined or inspiring champion.

ABOVE: Caged alone for years in an African
zoo, a male chimpanzee reaches to touch a
rare, understanding stranger: Jane Goodall.

PREVIOUS PAGES, LEFT: A camera crew films
Jane and Emmanuel Mtiti for the feature-film
documentary *Jane's Journey*.

PREVIOUS PAGES, RIGHT: Jane walks on stage to
the applause of 52,000 people at the Live Earth
concert at Giants Stadium in New Jersey.

"So, let us move forward with faith in ourselves, in our intelligence, in our indomitable spirit. Let us develop respect for all living things. Let us try to replace violence and intolerance with understanding and compassion. And love."

—JANE GOODALL

OPPOSITE: Young people, such as these children walking happily near the Ngamba Island Sanctuary in Uganda, are Jane's hope for the future.

FOLLOWING PAGES: Jane and orphaned chimpanzee Pasa gaze across the water from the shore of Ngamba Island Sanctuary.

# ABOUT THE JANE GOODALL INSTITUTE

FOUNDED IN 1977, THE JANE GOODALL INSTITUTE CONTINUES DR. GOODALL'S

PIONEERING RESEARCH ON CHIMPANZEE BEHAVIOR—RESEARCH THAT TRANSFORMED

SCIENTIFIC PERCEPTIONS OF THE RELATIONSHIP BETWEEN HUMANS AND ANIMALS.

TODAY, IT IS A GLOBAL LEADER IN THE EFFORT TO PROTECT CHIMPANZEES AND

THEIR HABITATS. IT ALSO IS WIDELY RECOGNIZED FOR ESTABLISHING INNOVATIVE

COMMUNITY-CENTERED CONSERVATION AND DEVELOPMENT PROGRAMS IN AFRICA,

AND JANE GOODALL'S ROOTS & SHOOTS GLOBAL ENVIRONMENT AND HUMANITARIAN

YOUTH PROGRAM. FOR MORE INFORMATION, PLEASE VISIT WWW.JANEGOODALL.ORG

AND WWW.ROOTSANDSHOOTS.ORG.

ABOVE: Today the Gombe Stream Research Centre is run by a dedicated field staff that includes many native Tanzanians and hosts a regular stream of visiting researchers. Here Jane is photographed in Gombe with several members of her team (left to right): Matendo Msafiri, sample collector, health monitoring; Dr. Anthony Collins, the director of baboon research; Sufi Hamisi, a field assistant for baboon research; Gabo Paulo, head field assistant for chimpanzee research; Issa Salala, the vice-head field assistant for chimpanzee research; Jumanne Kikwale, the manager of the Gombe Stream Research Centre; Dr. Shadrack Kamenya, the director of conservation science at JGI-Tanzania; and Dr. Iddi Lipende, a veterinarian for the health-monitoring program, GSRC.

PREVIOUS PAGES: Jane's travels take her around the world more than three hundred days of the year.

# Offices Worldwide

Something like a modern-day Johnny Appleseed, Dr. Goodall sows seeds of change as she travels the world. Many of the people Jane inspires during her talks and lectures go on to offer their time and energy to Jane's efforts and become integral members of the global Jane Goodall Institute family. There are more than twenty-five JGIs around the world, all committed to supporting Jane's global activities, especially the growth of Jane Goodall's Roots & Shoots program. Many local offices take on special missions as well. JGI-Italy, for example, has supported the Sanganigwa Children's Home, an orphanage in Kigoma (near Gombe), for decades; while JGI-Uganda does many school-based environmental-education activities. See below for a complete listing of JGIs around the world. Questions regarding any of the JGI offices without its own website may be directed to the United States office.

**JGI-AUSTRALIA**
www.janegoodall.org.au

**JGI-AUSTRIA**
www.janegoodall.at

**JGI-BELGIUM**
www.janegoodall.be

**JGI-CANADA**
www.janegoodall.ca

**JGI-CHINA:**

Roots & Shoots Beijing
www.jgichina.org

Roots & Shoots Chengdu
http://cdgyy.org

Roots & Shoots Shanghai
www.jgi-shanghai.org

**JGI-REPUBLIC OF CONGO**

**JGI-FRANCE**
www.janegoodall.fr

**JGI-GABON**
www.janegoodall.fr

**JGI-GERMANY**

**JGI-HONG KONG**
www.janegoodall.org.hk

**JGI-HUNGARY**
www.janegoodall.hu

**JGI-ITALY**
www.janegoodall-italia.org

**JGI-JAPAN**
www.jgi-japan.org

**JGI-KENYA**

**ROOTS & SHOOTS LATIN AMERICA AND CARIBBEAN**
web.mac.com/rickasselta/iweb/site/welcome.html

**JGI-NETHERLANDS**
www.janegoodall.nl

**JGI-SINGAPORE**
www.janegoodall.org.sg

**JGI-SOUTH AFRICA**
www.janegoodall.co.za
www.rootsandshoots.org.za

**JGI-SPAIN**
www.janegoodall.es

**JGI-SWEDEN**
www.swedenchimp.se/jgi-sweden.html

**JGI-SWITZERLAND**
www.janegoodall.ch

**JGI-TAIWAN**
www.goodall.org.tw

**JGI-TANZANIA**

**JGI-UGANDA**
www.jgiuganda.org

**JGI-UNITED KINGDOM**
www.janegoodall.org.uk

**JGI-USA**
www.janegoodall.org
www.rootsandshoots.org

# Milestones (1960–2010)

**JULY 14, 1960**
Jane Goodall begins her study in Gombe Stream Chimpanzee Reserve, accompanied by her mother, Vanne.

**OCTOBER 30, 1960**
Chimpanzees are first seen eating meat.

**NOVEMBER 1960**
Jane observes David Greybeard using a grass stem to fish for termites. It is the first scientific documentation of chimpanzee tool use.

**JANUARY 31, 1961**
Jane first observes the chimpanzees performing a "rain dance."

**MARCH 13, 1961**
Jane receives a grant of $1,400 from the National Geographic Society's Committee for Research and Exploration. It is the first of twenty-seven grants that she will receive from NGS.

**SUMMER 1961**
David Greybeard is the first chimp to explore Jane's camp.

**1962**
Jane enters the University of Cambridge as a Ph.D. candidate.

**1962 & 1964**
Jane receives the Franklin Burr Award from the National Geographic Society for her contribution to science.

**AUGUST 1963**
Jane's first article, "My Life Among Wild Chimpanzees," is published in *National Geographic* magazine.

**MARCH 28, 1964**
Jane Goodall marries wildlife filmmaker and photographer Hugo van Lawick.

**MARCH 1964**
Flo gives birth to Flint, whose upbringing provided a chance to observe chimpanzee parenting from its beginning.

**SPRING 1964**
Mike gains dominance in the community, and defeats Goliath, by charging with noisy kerosene cans and intimidating the other chimps.

**SUMMER 1964**
Evered is the first chimp seen using chewed leaves as a sponge to soak up water. This is another tool that is frequently used by the chimpanzees in Gombe.

**SUMMER 1964**
Chimpanzees are first seen using leaves to clean themselves and wipe wounds.

**1965**
The Gombe Stream Research Centre is founded.

**1965**
Jane Goodall receives her Ph.D. in ethology, the study of animal behavior, from the University of Cambridge. She is the eighth person at the university to be awarded a Ph.D. without first receiving an undergraduate degree.

**DECEMBER 1965**
Jane's second article, "New Discoveries Among Africa's Chimpanzees," with photographs by Hugo van Lawick, appears in *National Geographic* magazine.

**DECEMBER 1965**
*Miss Goodall and the Wild Chimpanzees*, an hour-long program produced by the National Geographic Society, appears on national television in the United States.

**1966**
Fifteen Kasekela chimps are afflicted with polio. In the end, six die from the disease and the survivors have afflictions that leave them disabled for the rest of their lives.

**1966**
Study on Gombe's baboons begins.

**1967**
The Gombe Stream Chimpanzee Reserve becomes Gombe National Park.

**MARCH 4, 1967**
Jane Goodall's son, Hugo Eric Louis van Lawick, nicknamed Grub, is born.

**1968**
Hilali Matama, Jane's first official field assistant, is hired at the Gombe Stream Research Centre.

**1968**
David Greybeard dies.

**1970**
Jane's first children's book, *Grub, the Bush Baby*, is published.

**1971**
Humphrey defeats Mike and takes over as the alpha male of the Kasekela group.

**1971**
Jane's book *In the Shadow of Man* is published. It became an instant bestseller.

**MAY 1971**
Fifi, daughter of Flo, gives birth to Freud, the first of seven offspring she raised to adulthood.

**1972**
Humphrey's twenty-month reign as alpha male is ended by Figan.

**AUGUST 22, 1972**
Flo, the fearless and loving mother, dies of old age.

**SEPTEMBER 15, 1972**
Flint, Flo's eight-year-old offspring, becomes depressed and dies soon after the death of his mother.

**EARLY 1974**
The start of the "four-year war" at Gombe, the first record of long-term "warfare" among chimpanzees. The original Kasekela group was divided and members of the new Kahama group were systematically annihilated.

**1975**
Jane marries her second husband, the Hon. Derek Bryceson, a member of Tanzanian parliament and the director of Tanzanian national parks.

**1977**
The Jane Goodall Institute for Wildlife Research, Education, and Conservation is established in San Francisco by Jane Goodall, Ranieri di San Faustino, and Genevieve di San Faustino.

**OCTOBER 21, 1977**
Melissa gives birth to twins Gyre and Gimble. Only Gimble survives.

**MAY 1979**
Jane reports on new discoveries in her article "Life and Death at Gombe," for *National Geographic*.

**1980**
Jane receives the Order of the Golden Ark, the World Wildlife award for conservation, presented to her by Prince Bernhard of the Netherlands.

**1982**
After a three-year struggle with Figan, Goblin finally gains control as alpha male.

## FEBRUARY 10, 1982
Passion dies after falling ill. Passion, along with her daughter Pom, had killed between five and ten newborn chimpanzees for food.

## JUNE 1982
Gremlin, daughter of Melissa, gives birth to Getty, only the second birth to be observed in twenty-two years of research.

## 1984
Jane's second National Geographic special, *Among the Wild Chimpanzees*, airs on national television.

## SPRING 1984
The ChimpanZoo project is conceived with the goals of recording behavior of captive chimpanzees and promoting stimulating environments for chimpanzees and other primates.

## OCTOBER 1986
Melissa, mother of Goblin, Gremlin, and Gimble, dies.

## 1986
Jane publishes *The Chimpanzees of Gombe: Patterns of Behavior*, a comprehensive scholarly analysis of chimpanzee behavior.

## NOVEMBER 1986
At a scientific conference in Chicago, Jane is shocked to learn of the widespread habitat destruction across Africa. She leaves the conference knowing she must leave Gombe behind and work to save the chimpanzees.

## MARCH 1987
An outbreak of pneumonia afflicts the chimpanzees, killing nine. It is the worst epidemic since the polio outbreak in 1966.

## 1987
After three-year-old Mel's mother died of pneumonia, he was "adopted" by an adolescent male, Spindle—the first time that a non-related chimp is observed to adopt an orphaned youngster.

## MAY 25, 1988
JGI-UK is established in London.

## SPRING 1990
Wilkie defeats Goblin, whose reign lasted nine years, to become alpha male at Gombe.

## JULY 1990
JGI-Tanzania is launched in conjunction with Jane Goodall's Gombe 30 celebration, observing thirty years since she first began her research in Tanzania.

## 1990
*Chimps, So Like Us*, an HBO documentary, is produced. The film is nominated for an Academy Award.

## 1990
*Through a Window*, Jane's fifth book, is published.

## 1990
Jane receives the Kyoto Prize in basic science, the Japanese equivalent of the Nobel Prize.

## FEBRUARY 1991
Jane and sixteen Tanzanian students found Jane Goodall's Roots & Shoots in Dar es Salaam, Tanzania.

## 1992
The Tchimpounga Chimpanzee Rehabilitation Center opens. It was initially funded by Conoco Inc.

## 1993
Freud, eldest son of Fifi, defeats Wilkie to become alpha male of Gombe's Kasekela group.

## FEBRUARY 14, 1993
Videographer Bill Wallauer records the first video footage of a chimpanzee birth in the wild: Gremlin giving birth to her daughter Gaia.

## 1994
After a Mitumba chimp joined the Kasekela group, Flossi, daughter of Fifi, suddenly begins using the Mitumba technique of catching carpenter ants with twigs. It is the first observation of technology transfer from one community of chimpanzees to another.

## 1994
JGI founds TACARE.

## 1995
Jane receives the National Geographic Society's Hubbard Medal for distinction in exploration, discovery, and research. The award is presented to her by Vice President Al Gore.

## FEBRUARY 1995
Rafiki, in Gombe's Mitumba community, gives birth to twins, whom Jane names Roots and Shoots. They are only the second set of twins observed at Gombe.

## AUGUST 1995
The Jane Goodall Institute's Center for Primate Studies is founded at the University of Minnesota.

## 1996
Pneumonia strikes the Mitumba group, killing about one-third of the population. Rafiki and her young twins, Roots and Shoots, are three of the victims.

## 1996
Jane receives the Tanzanian Kilimanjaro Medal, presented by President Mwinyi, for her contributions to wildlife conservation.

## SUMMER 1997
Mange, a skin disease, infiltrates the Kasekela community, hitting hardest on the nursing females and their infants. Fifi loses her infant son, Fred. Also affected are Freud, Prof, Goblin, and Beethoven.

## OCTOBER 2, 1997
Frodo overthrows his ailing brother Freud as alpha male of the Kasekela group.

## JULY 1998
Gremlin gives birth to Gombe's newest set of twins, Golden and Glitta. Fifi gives birth to her third daughter, Flirt.

## 1999
Jane's eighth book, *Reason for Hope: A Spiritual Journey*, is published and instantly becomes a *New York Times* bestseller.

## 2001
Jane is awarded the Gandhi-King Award for Non-Violence.

## APRIL 16, 2002
United Nations Secretary-General Kofi Annan appoints Jane to serve as a United Nations Messenger of Peace. Secretary-General Ban Ki-moon reappoints Jane in June 2007.

## 2003
Jane receives the Prince of Asturias Award for Technical and Scientific Research. She also receives the Benjamin Franklin Medal in Life Science that same year.

## 2003
JGI begins work with Discovery Communications' Animal Planet and produces five television movies over the course of six years.

## FEBRUARY 20, 2004
Jane is made a Dame of the British Empire.

## 2004
Fifi disappears with daughter Furaha, born in 2002, and is assumed dead.

## 2004
Goblin dies after an illness.

## 2006
Jane receives the French Legion of Honor, presented by Prime Minister Dominique de Villepin. She also receives the UNESCO Gold Medal that same year.

## 2009
Jane publishes *Hope for Animals and Their World: How Endangered Species Are Being Rescued from the Brink*.

## 2010
The film *Jane's Journey* opens.

## 2010
The Jane Goodall Institute commemorates the fiftieth anniversary of Jane's arrival at Gombe with a worldwide celebration.

# Acknowledgments

It goes without saying that it has taken the hard work of many people to assemble the material for this book. First, grateful thanks to Jennifer Lindsey, who wrote *Jane Goodall: 40 Years at Gombe*, the book on which this one is based. Thanks to Leslie Stoker, senior vice president and publisher at Stewart, Tabori & Chang, for accepting the challenge to revise the book and get it ready for our anniversary, and to Ann Stratton for her day-to-day dedication to the task.

Much gratitude to all the photographers who contributed their talents to help tell the Gombe story in such a compelling way.

I am also so appreciative of all the people on the JGI team—the JGI office in Arlington, Virginia; the Jane Goodall Institute's Center for Primate Studies at the University of Minnesota; and those who made contributions or checked for accuracy from the JGI offices in Tanzania—all worked hard to ensure the information in this book was factually correct and that new data was included. Mary Paris spent countless hours researching photos and editing the images that appear on these pages. Her creative input throughout the development of this book was invaluable. And just when it seemed as if we would never be able to get everything together on time, Jackie Conciatore, who is a wonderful writer, came to our rescue, working evenings and weekends to ensure we met our deadline.

Two people in particular worked extraordinarily hard to ensure that *Jane Goodall: 50 Years at Gombe* was finished in a professional way and would be published in time for our fiftieth-anniversary celebrations: Christin Jones, a valued staff member of the Office of the Founder-Global, and Nona Gandelman, a true friend of JGI and now my agent. Christin worked closely with Nona, ensuring that all members of the team provided the information necessary in a timely way. Without Nona, this book might not have happened: It was she who worked closely with the publisher and ensured that the chapters flowed smoothly. And it was she who—with Christin's help—bullied me into working on the manuscript, making the necessary additions, and writing the new introduction.

Indeed, the production of this book has been a team effort, a triumph of cooperation. Thank you all so very much.

*Jane Goodall*

ABOVE: Jane first saw La Vielle at the Pointe Noire zoo in Congo. Her cage door was broken, and she was starving, but she would not step out of her cage, even for food. Jane was able to get her moved to the Tchimpounga Sanctuary. She was so traumatized that once there it took two years before she ventured out of her night quarters onto the grass. Eventually a garden was created at Tchimpounga for her and old Gregoire. Since Gregoire's death in 2008, La Vielle has adopted one of the newly arrived orphans. They are inseparable.

RIGHT: As today's visitors to Gombe depart the park, they see this sign.

# Index

**All pages in *italics* refer to illustrations.**